Stochastic Models for Geodesy and Geoinformation Science

Stochastic Models for Geodesy and Geoinformation Science

Editor

Frank Neitzel

MDPI • Basel • Beijing • Wuhan • Barcelona • Belgrade • Manchester • Tokyo • Cluj • Tianjin

Editor
Frank Neitzel
Institute of Geodesy and
Geoinformation Science,
Technische Universität Berlin
Germany

Editorial Office
MDPI
St. Alban-Anlage 66
4052 Basel, Switzerland

This is a reprint of articles from the Special Issue published online in the open access journal *Mathematics* (ISSN 2227-7390) (available at: https://www.mdpi.com/journal/mathematics/special_issues/Stochastic_Models_Geodesy_Geoinformation_Science).

For citation purposes, cite each article independently as indicated on the article page online and as indicated below:

LastName, A.A.; LastName, B.B.; LastName, C.C. Article Title. *Journal Name* **Year**, *Volume Number*, Page Range.

ISBN 978-3-03943-981-2 (Hbk)
ISBN 978-3-03943-982-9 (PDF)

© 2021 by the authors. Articles in this book are Open Access and distributed under the Creative Commons Attribution (CC BY) license, which allows users to download, copy and build upon published articles, as long as the author and publisher are properly credited, which ensures maximum dissemination and a wider impact of our publications.

The book as a whole is distributed by MDPI under the terms and conditions of the Creative Commons license CC BY-NC-ND.

Contents

About the Editor vii

Preface to "Stochastic Models for Geodesy and Geoinformation Science" ix

Burkhard Schaffrin
Total Least-Squares Collocation: An Optimal Estimation Technique for the EIV-Model with Prior Information
Reprinted from: *Mathematics* **2020**, *8*, 971, doi:10.3390/math8060971 1

Georgios Malissiovas, Sven Weisbrich, Frank Neitzel and Svetozar Petrovic
Weighted Total Least Squares (WTLS) Solutions for Straight Line Fitting to 3D Point Data
Reprinted from: *Mathematics* **2020**, *8*, 1450, doi:10.3390/math8091450 11

Wolfgang Niemeier and Dieter Tengen
Stochastic Properties of Confidence Ellipsoids after Least Squares Adjustment, Derived from GUM Analysis and Monte Carlo Simulations
Reprinted from: *Mathematics* **2020**, *8*, 1318, doi:10.3390/math8081318 31

Rüdiger Lehmann, Michael Lösler and Frank Neitzel
Mean Shift versus Variance Inflation Approach for Outlier Detection—A Comparative Study
Reprinted from: *Mathematics* **2020**, *8*, 991, doi:10.3390/math8060991 49

Till Schubert, Johannes Korte, Jan Martin Brockmann and Wolf-Dieter Schuh
A Generic Approach to Covariance Function Estimation Using ARMA-Models
Reprinted from: *Mathematics* **2020**, *8*, 591, doi:10.3390/math8040591 71

Pakize Küreç Nehbit, Robert Heinkelmann, Harald Schuh, Susanne Glaser, Susanne Lunz, Nicat Mammadaliyev, Kyriakos Balidakis, Haluk Konak and Emine Tanır Kayıkçı
Evaluation of VLBI Observations with Sensitivity and Robustness Analyses
Reprinted from: *Mathematics* **2020**, *8*, 939, doi:10.3390/math8060939 91

Yumiao Tian, Maorong Ge and Frank Neitzel
Variance Reduction of Sequential Monte Carlo Approach for GNSS Phase Bias Estimation
Reprinted from: *Mathematics* **2020**, *8*, 522, doi:10.3390/math8040522 107

Xinggang Zhang, Pan Li, Rui Tu, Xiaochun Lu, Maorong Ge and Harald Schuh
Automatic Calibration of Process Noise Matrix and Measurement Noise Covariance for Multi-GNSS Precise Point Positioning
Reprinted from: *Mathematics* **2020**, *8*, 502, doi:10.3390/math8040502 123

Gabriel Kerekes and Volker Schwieger
Elementary Error Model Applied to Terrestrial Laser Scanning Measurements: Study Case Arch Dam Kops
Reprinted from: *Mathematics* **2020**, *8*, 593, doi:10.3390/math8040593 143

Gaël Kermarrec
On Estimating the Hurst Parameter from Least-Squares Residuals. Case Study: Correlated Terrestrial Laser Scanner Range Noise
Reprinted from: *Mathematics* **2020**, *8*, 674, doi:10.3390/math8050674 165

About the Editor

Frank Neitzel is Full Professor at Technische Universität Berlin (TU Berlin). He is Managing Director of the Institute of Geodesy and Geoinformation Science and Head of the Chair of Geodesy and Adjustment Theory. Prior to his work at TU Berlin he was Professor of Engineering Geodesy at the University of Applied Sciences Mainz, Germany. Dr. Neitzel was awarded a Feodor Lynen Research Fellowship of the Alexander von Humboldt Foundation (Germany) for a research stay at the School of Earth Sciences at The Ohio State University (USA). Dr. Neitzel's research interests include rigorous solutions for total least squares problems considering singular variance–covariance matrices, deformation analysis, structural health monitoring, and surface approximation. At TU Berlin, his laboratory for geodetic measurement and data analysis focuses especially on the use of terrestrial laser scanners, image-based measurement systems, and accelerometers for the monitoring of built and natural objects. This includes the monitoring of deformations as well as investigations of the vibration characteristics. His research group developed an approach for automated deformation analysis using point clouds captured by terrestrial laser scanners and an intensity-based stochastic model for laser scanner data. His interdisciplinary work focuses on measurement- and model-based structural analysis, combining measurement-based approaches from geodesy with model-based approaches from civil engineering. Dr. Neitzel has worked on research projects funded by the German Research Foundation (DFG), the Federal Ministry of Education and Research (BMBF), and the Federal Ministry for Economic Affairs and Energy (BMWi), among others. He is the author of more than 100 publications in peer-reviewed journals, book chapters, and conference proceedings.

Preface to "Stochastic Models for Geodesy and Geoinformation Science"

In geodesy and geoinformation science, as well as in many other technical disciplines, it is often not possible to directly determine the desired target quantities, for example, the 3D Cartesian coordinates of an object. Therefore, the unknown parameters must be linked with measured values, for instance, directions, angles, and distances, by a mathematical model. This consists of two fundamental components—the functional and the stochastic models. The functional model describes the geometrical–physical relationship between the measured values and the unknown parameters. This relationship is sufficiently well known for most applications.

With regard to stochastic models in geodesy and geoinformation science, two problem domains of fundamental importance arise:

1. How can stochastic models be set up as realistically as possible for the various geodetic observation methods and sensor systems, such as very-long-baseline interferometry (VLBI), global navigation satellite systems (GNSS), terrestrial laser scanners, and multisensor systems?

2. How can the stochastic information be adequately considered in appropriate least squares adjustment models?

Further questions include the interpretation of the stochastic properties of the computed target values for quality assessment of the results in terms of precision and reliability and for the detection of outliers in the input data (measurements).

In this Special Issue, current research results on these general questions are presented in ten peer-reviewed articles. The basic findings can be applied to all technical scientific fields where measurements are used for the determination of parameters to describe geometric or physical phenomena.

Frank Neitzel
Editor

Article

Total Least-Squares Collocation: An Optimal Estimation Technique for the EIV-Model with Prior Information

Burkhard Schaffrin

Geodetic Science Program, School of Earth Sciences, The Ohio State University, Columbus, OH 43210, USA; schaffrin.1@osu.edu

Received: 3 May 2020; Accepted: 1 June 2020; Published: 13 June 2020

Abstract: In regression analysis, oftentimes a linear (or linearized) Gauss-Markov Model (GMM) is used to describe the relationship between certain unknown parameters and measurements taken to learn about them. As soon as there are more than enough data collected to determine a unique solution for the parameters, an estimation technique needs to be applied such as 'Least-Squares adjustment', for instance, which turns out to be optimal under a wide range of criteria. In this context, the matrix connecting the parameters with the observations is considered fully known, and the parameter vector is considered fully unknown. This, however, is not always the reality. Therefore, two modifications of the GMM have been considered, in particular. First, 'stochastic prior information' (p. i.) was added on the parameters, thereby creating the – still linear – Random Effects Model (REM) where the optimal determination of the parameters (random effects) is based on 'Least Squares collocation', showing higher precision as long as the p. i. was adequate (Wallace test). Secondly, the coefficient matrix was allowed to contain observed elements, thus leading to the – now nonlinear – Errors-In-Variables (EIV) Model. If not using iterative linearization, the optimal estimates for the parameters would be obtained by 'Total Least Squares adjustment' and with generally lower, but perhaps more realistic precision. Here the two concepts are combined, thus leading to the (nonlinear) 'EIV-Model with p. i.', where an optimal estimation (resp. prediction) technique is developed under the name of 'Total Least-Squares collocation'. At this stage, however, the covariance matrix of the data matrix – in vector form – is still being assumed to show a Kronecker product structure.

Keywords: Errors-In-Variables Model; Total Least-Squares; prior information; collocation vs. adjustment

---

1. Introduction

Over the last 50 years or so, the (linearized) Gauss-Markov Model (GMM) as standard model for the estimation of parameters from collected observation [1,2] has been refined in a number of ways. Two of these will be considered in more detail, namely

- the GMM after *strengthening* the parameters through the introduction of "stochastic prior information". The relevant model will be the "Random Effects Model (REM)", and the resulting estimation technique has become known as "*least-squares collocation*" [3,4].
- the GMM after *weakening* the coefficient matrix through the replacement of fixed entries by observed data, resulting in the (nonlinear) Errors-In-Variables (EIV) Model. When *nonlinear* normal equations are formed and subsequently solved by iteration, the resulting estimation technique has been termed "*Total Least-Squares (TLS) estimation*" [5–7]. The alternative approach, based on iteratively linearizing the EIV-Model, will lead to identical estimates of the parameters [8].

After a compact review and comparison of several key formulas (parameter estimates, residuals, variance component estimate, etc.) within the above three models, a new model will be introduced that allows the strengthening of the parameters and the weakening of the coefficient matrix at the same time. The corresponding estimation technique will be called "TLS collocation" and follows essentially the outline that had first been presented by this author in June 2009 at the Intl. Workshop on Matrices and Statistics in Smolenice Castle (Slovakia); cf. Schaffrin [9].

Since then, further computational progress has been made, e.g. Snow and Schaffrin [10]; but several open questions remain that need to be addressed elsewhere. These include issues related to a rigorous error propagation of the "TLS collocation" results; progress may be achieved here along similar lines as in Snow and Schaffrin [11], but is beyond the scope of the present paper.

For a general overview of the participating models, the following diagram (Figure 1) may be helpful.

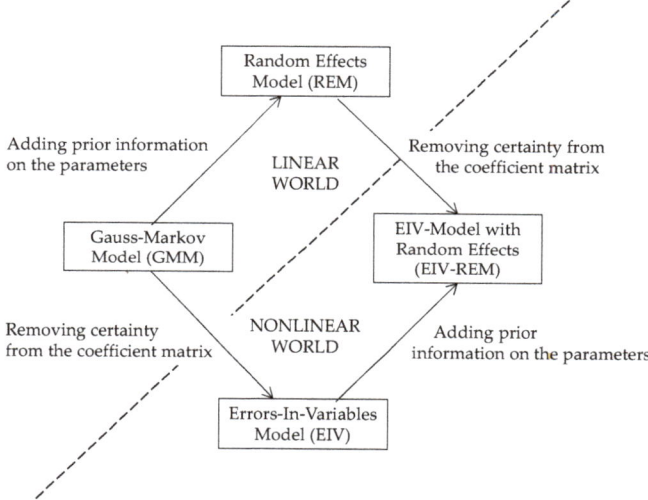

Figure 1. Model Diagram.

The most informative model defines the top position, and the least informative model is at the bottom. The new model can be formed at the intermediate level (like the GMM), but belongs to the "nonlinear world" where *nonlinear* normal equations need to be formed and subsequently solved by iteration.

2. A Compact Review of the "Linear World"

2.1. The (linearized) Gauss-Markov Model (GMM)

Let the Gauss-Markov Model be defined by

$$y = \underset{n \times m}{A}\, \xi + e\,,\ q := \text{rk}\, A \leq \min\{m, n\}\,,\ e \sim (0, \sigma_0^2 I_n)\,, \tag{1}$$

possibly after linearization and homogeneization; cf. Koch [2], or Grafarend and Schaffrin [12] among many others.

Here, y denotes the $n \times 1$ vector of (incremental) observations,

A the $n \times m$ matrix of coefficients (given),
ξ the $m \times 1$ vector of (incremental) parameters (unknown),

e the $n \times 1$ vector of random errors (unknown) with expectation $E\{e\} = 0$ while the dispersion matrix $D\{e\} = \sigma_0^2 I_n$ is split into the (unknown) factor

σ_0^2 as variance component (unit-free) and

I_n as (homogenized) $n \times n$ symmetric and positive-definite "cofactor matrix" whose inverse is better known as "weight matrix" P; here $P = I_n$ for the sake of simplicity.

Now, it is well known that the Least-Squares Solution (LESS) is based on the principle

$$e^T e = \min \quad \text{s.t. } e = y - A\xi \tag{2}$$

which leads to the "normal equations"

$$N\hat{\xi} = c \quad \text{for} \quad [N, c] := A^T[A, y]. \tag{3}$$

Depending on rk N, the rank of the matrix N, the LESS may turn out *uniquely* as

$$\hat{\xi}_{LESS} = N^{-1} c \quad \text{iff} \quad \text{rk } N = \text{rk } A = q = m; \tag{4}$$

or it may belong to a *solution (hyper)space* that can be characterized by certain generalized inverses of N, namely

$$\begin{aligned}\hat{\xi}_{LESS} &\in \{N^- c | N N^- N = N\} = \\ &= \{N_{rs}^- c | N N_{rs}^- N = N, \ N_{rs}^- N N_{rs}^- = N_{rs}^- = (N_{rs}^-)^T\};\end{aligned} \tag{5}$$

where N_{rs}^- denotes an arbitrary *reflexive symmetric g-inverse* of N (including the "pseudo-inverse" N^+),

$$\text{iff} \quad \text{rk } N = \text{rk } A = q < m. \tag{6}$$

In the latter case (6), any LESS will be *biased* with

$$\text{bias}(\hat{\xi}_{LESS}) = E\{\hat{\xi}_{LESS} - \xi\} \in \{-(I_m - N^- N)\xi\} = \{-(I_m - N_{rs}^- N)\xi\}, \tag{7}$$

its dispersion matrix will be

$$D\{\hat{\xi}_{LESS}\} = \sigma_0^2 N^- N(N^-)^T \in \{\sigma_0^2 N_{rs}^- | N N_{rs}^- N = N, \ N_{rs}^- N N_{rs}^- = N_{rs}^- = (N_{rs}^-)^T\}, \tag{8}$$

and its Mean Squared Error (MSE) matrix, therefore,

$$\text{MSE}\{\hat{\xi}_{LESS}\} \in \{\sigma_0^2 [N_{rs}^- + (I_m - N_{rs}^- N)(\xi \sigma_0^{-2} \xi^T)(I_m - N_{rs}^- N)^T]\}. \tag{9}$$

In contrast to the choice of LESS in case (6), the *residual vector* will be *unique*; for any $\hat{\xi}_{LESS}$:

$$\tilde{e}_{LESS} = y - A\hat{\xi}_{LESS} \sim (0, \sigma_0^2 (I_n - A N_{rs}^- A^T) = D\{y\} - D\{A\hat{\xi}_{LESS}\}). \tag{10}$$

Hence, the *optimal variance component estimate* will also be *unique*:

$$\hat{\sigma}_0^2 = (n - q)^{-1} \tilde{e}_{LESS}^T \cdot \tilde{e}_{LESS} = (n - q)^{-1} (y^T y - c^T \hat{\xi}_{LESS}), \tag{11}$$

$$E\{\hat{\sigma}_0^2\} = \sigma_0^2, \ D\{\hat{\sigma}_0^2\} \approx 2(\sigma_0^2)^2/(n - q) \quad \text{under quasi-normality.} \tag{12}$$

For the corresponding formulas in the full-rank case (4), the reflexive symmetric g-inverse N_{rs}^- simply needs to be replaced by the regular inverse N^{-1}, thereby showing that the LESS turns into an unbiased estimate of ξ while its **MSE**-matrix coincides with its dispersion matrix (accuracy precision).

2.2. The Random Effects Model (REM)

Now, additional prior information (p. i.) is introduced to *strengthen* the parameter vector within the GMM (1). Based on Harville's [13] notation of a $m \times 1$ vector $\underset{\sim}{0}$ of so-called *"pseudo-observations"*, the new $m \times 1$ vector

$$x := \xi - \underset{\sim}{0} \text{ of "random effects" (unknown) is formed with} \tag{13}$$

$$E\{x\} = \beta_0 \text{ as given } m \times 1 \text{ vector of p. i. and} \tag{14}$$

$$D\{x\} = \sigma_0^2 Q_0 \text{ as } n \times n \text{ positive-(semi)definite dispersion matrix of the p. i.} \tag{15}$$

Consequently, the Random Effects Model (REM) can be stated as

$$\underset{\sim}{y} := y - A \cdot \underset{\sim}{0} = \underset{n \times m}{A} x + e, \ q := \text{rk } A \leq \min\{m, n\}, \ \underset{m \times 1}{\beta_0} := x + e_0, \tag{16}$$

$$\begin{bmatrix} e \\ e_0 \end{bmatrix} \sim \left(\begin{bmatrix} 0 \\ 0 \end{bmatrix}, \sigma_0^2 \begin{bmatrix} I_n & 0 \\ 0 & Q_0 \end{bmatrix} \right), \ \underset{m \times m}{Q_0} \text{ symmetric and nnd (non-negative definite)}. \tag{17}$$

If $P_0 := Q_0^{-1}$ exists, the estimation/prediction of x may now be based on the principle

$$e^T e + e_0^T P_0 e_0 = \min \text{ s.t. } e = \underset{\sim}{y} - Ax, \ e_0 = \beta_0 - x, \tag{18}$$

which leads to the "normal equations" for the "Least-Squares Collocation (LSC)" solution

$$(N + P_0) \cdot \tilde{x} = \underset{\sim}{c} + P_0 \cdot \beta_0 \tag{19}$$

which is *unique* even in the rank-deficient case (6).

If Q_0 is singular, but $e_0 \in \mathfrak{R}(Q_0)$ with probability 1, then there must exist an $m \times 1$ vector v^0 with

$$Q_0 v^0 = -e_0 = x - \beta_0 \text{ with probability 1 (a.s. = almost surely).} \tag{20}$$

Thus, the principle (18) may be *equivalently* replaced by

$$e^T e + (v^0)^T Q_0 v^0 = \min \text{ s.t. } e = \underset{\sim}{y} - Ax \text{ and (20)}, \tag{21}$$

which generates the *LSC solution* uniquely from the "modified normal equations"

$$\boxed{(I_m + Q_0 N) \cdot \tilde{x}_{\text{LSC}} = \beta_0 + Q_0 \underset{\sim}{c}} \text{ for } \underset{\sim}{c} := A^T \underset{\sim}{y} \tag{22}$$

or, alternatively, via the "update formula"

$$\boxed{\tilde{x}_{\text{LSC}} = \beta_0 + Q_0 (I_m + NQ_0)^{-1} (\underset{\sim}{c} - N\beta_0)} \tag{23}$$

which exhibits the "*weak (local) unbiasedness*" of $\tilde{x}_{\text{LSC}}$ via

$$E\{\tilde{x}_{\text{LSC}}\} = \beta_0 + Q_0 (I_m + NQ_0)^{-1} (E\{\underset{\sim}{c}\} - N\beta_0) = \beta_0 + 0 = E\{x\}. \tag{24}$$

Consequently, the **MSE**-matrix of $\tilde{x}_{\text{LSC}}$ can be obtained from

$$\textbf{MSE}\{\tilde{x}_{\text{LSC}}\} = D\{\tilde{x}_{\text{LSC}} - x\} = \sigma_0^2 Q_0 (I_m + NQ_0)^{-1} = \tag{25}$$

$$= \sigma_0^2 \left(N + P_0\right)^{-1} \text{ if } P_0 := Q_0^{-1} \text{ exists,} \tag{26}$$

whereas the dispersion matrix of $\tilde{x}_{LSC}$ itself is of *no relevance* here. It holds:

$$D\{\tilde{x}_{LSC}\} = D\{x\} - D\{x - \tilde{x}_{LSC}\} = \sigma_0^2 Q_0 - \sigma_0^2 Q_0 (I_m + NQ_0)^{-1}. \tag{27}$$

Again, *both residual vectors* are uniquely determined from

$$(\tilde{e}_0)_{LSC} = \beta_0 - \tilde{x}_{LSC} = -Q_0 \, \hat{v}_{LSC}^0 \text{ for} \tag{28}$$

$$\hat{v}_{LSC}^0 := (I_m + NQ_0)^{-1}(\underline{c} - N\beta_0), \text{ and} \tag{29}$$

$$\tilde{e}_{LSC} = \underline{y} - A\,\tilde{x}_{LSC} = [I_m - A\,Q_0(I_m + NQ_0)^{-1}A^T]\,(\underline{y} - A\beta_0) \tag{30}$$

with the control formula

$$\boxed{A^T \tilde{e}_{LSC} = \hat{v}_{LSC}^0}. \tag{31}$$

An optimal estimate of the *variance component* may now be obtained from

$$\begin{aligned}(\hat{\sigma}_0^2)_{LSC} &= n^{-1} \cdot (\tilde{e}_{LSC}^T \cdot \tilde{e}_{LSC} + (\hat{v}_{LSC}^0)^T Q_0 \, \hat{v}_{LSC}^0) = \\ &= n^{-1} \cdot (\underline{y}^T \underline{y} - \underline{c}^T \tilde{x}_{LSC} - \beta_0^T \hat{v}_{LSC}^0).\end{aligned} \tag{32}$$

3. An Extension into the "Nonlinear World"

3.1. The Errors-In-Variables (EIV) Model

In this scenario, the Gauss-Markov Model (GMM) is further *weakened* by allowing some or all of the entries in the coefficient matrix A to be observed. So, after introducing a corresponding $n \times m$ matrix of unknown random errors E_A, the original GMM (1) turns into the EIV-Model

$$\underline{y} = \underbrace{(A - E_A)}_{n \times m} \cdot \xi + \underline{e}, \; q := \text{rk } A = m < n, \; \underline{e} \sim (0, \sigma_0^2 I_n), \tag{33}$$

with the vectorized form of E_A being characterized through

$$\underline{e}_A \underset{nm \times 1}{:=} \text{vec } E_A \sim (0, \sigma_0^2(I_m \otimes I_n) = \sigma_0^2 I_{mn}), \; C\{\underline{e}, \underline{e}_A\} = 0 \; \text{(assumed)}. \tag{34}$$

Here, the vec operation transform a matrix into a vector by stacking all columns underneath each other while $\otimes$ denotes the *Kronecker-Zehfuss product* of matrices, defined by

$$\underset{p \times q}{G} \otimes \underset{r \times s}{H} = [g_{ij} \cdot H]_{pr \times qs} \text{ if } G = [g_{ij}]. \tag{35}$$

In particular, the following key formula holds true:

$$\text{vec}\,(A\,B\,C^T) = (C \otimes A) \cdot \text{vec } B \tag{36}$$

for matrices of suitable size. Note that, in (34), the choice $Q_A := I_{mn}$ is a very special one. In general, Q_A may turn out singular whenever some parts of the matrix A remain unobserved (i.e., nonrandom).

In any case, thanks to the term

$$E_A \cdot \xi = (\xi^T \otimes I_n)\,\underline{e}_A, \tag{37}$$

the model (33) needs to be treated in the *"nonlinear world"* even though the vector ξ may contain only incremental parameters. From now on, A is assumed to have *full column rank*, $\text{rk } A =: q = m$.

Following Schaffrin and Wieser [7], for instance, the *"Total-Least Squares Solution (TLSS)"* can be based on the principle

$$e^T e + e_A^T e_A = \min \text{ s.t. (33-34)}, \tag{38}$$

and the Lagrange function

$$\begin{aligned}\Phi(e, e_A, \xi, \lambda): &= e^T e + e_A^T e_A + 2\lambda^T(y - A\xi - e + E_A \xi) = \\ &= e^T e + e_A^T e_A + 2\lambda^T[y - A\xi - e + (\xi^T \otimes I_n) e_A]\end{aligned} \tag{39}$$

which needs to be made *stationary*. The necessary Euler-Lagrange conditions then read:

$$\frac{1}{2}\frac{\partial \Phi}{\partial e} = \tilde{e} - \hat{\lambda} \doteq 0 \Rightarrow \hat{\lambda} = \tilde{e} \tag{40}$$

$$\frac{1}{2}\frac{\partial \Phi}{\partial e_A} = \tilde{e}_A + (\hat{\xi}_{TLS} \otimes I_n)\hat{\lambda} \doteq 0 \Rightarrow \tilde{E}_A = -\hat{\lambda}\hat{\xi}_{TLS}^T \tag{41}$$

$$\frac{1}{2}\frac{\partial \Phi}{\partial \xi} = -(A - \tilde{E}_A)^T \hat{\lambda} \doteq 0 \Rightarrow A^T \hat{\lambda} = \tilde{E}_A^T \hat{\lambda} = -\hat{\xi}_{TLS} \cdot (\hat{\lambda}^T \hat{\lambda}) \tag{42}$$

$$\frac{1}{2}\frac{\partial \Phi}{\partial \lambda} = y - A\hat{\xi}_{TLS} - \tilde{e} + \tilde{E}_A \hat{\xi}_{TLS} \doteq 0 \Rightarrow y - A\hat{\xi}_{TLS} = \hat{\lambda}(1 + \hat{\xi}_{TLS}^T \hat{\xi}_{TLS}) \tag{43}$$

$$\Rightarrow c - N\hat{\xi}_{TLS} := A^T(y - A\hat{\xi}_{TLS}) = A^T \hat{\lambda}(1 + \hat{\xi}_{TLS}^T \hat{\xi}_{TLS}) = -\hat{\xi}_{TLS} \cdot \hat{v}_{TLS} \tag{44}$$

for

$$\hat{v}_{TLS} := (\hat{\lambda}^T \hat{\lambda}) \cdot (1 + \hat{\xi}_{TLS}^T \hat{\xi}_{TLS}) = \hat{\lambda}^T(y - A\hat{\xi}_{TLS}) = \tag{45}$$

$$= (1 + \hat{\xi}_{TLS}^T \hat{\xi}_{TLS})^{-1} \cdot (y - A\hat{\xi}_{TLS})^T(y - A\hat{\xi}_{TLS}) = \tag{46}$$

$$= (1 + \hat{\xi}_{TLS}^T \hat{\xi}_{TLS})^{-1} [y^T(y - A\hat{\xi}_{TLS}) - \hat{\xi}_{TLS}^T(c - N\hat{\xi}_{TLS})] \geq 0 \tag{47}$$

$$\Rightarrow (1 + \hat{\xi}_{TLS}^T \hat{\xi}_{TLS}) \cdot \hat{v}_{TLS} = y^T y - c^T \hat{\xi}_{TLS} + (\hat{\xi}_{TLS}^T \hat{\xi}_{TLS}) \cdot \hat{v}_{TLS}$$

$$\Rightarrow \boxed{\hat{v}_{TLS} = y^T y - c^T \hat{\xi}_{TLS}} \tag{48}$$

which needs to be solved in connection with the "modified normal equations" from (44), namely

$$\boxed{(N - \hat{v}_{TLS} I_m)\hat{\xi}_{TLS} = c}. \tag{49}$$

Due to the nonlinear nature of $\hat{\xi}_{TLS}$, it is not so easy to determine if it is an unbiased estimate, or how its MSE-matrix may exactly look like. First attempts of a rigorous error propagation have recently been undertaken by Amiri-Simkooei et al. [14] and by Schaffrin and Snow [15], but are beyond the scope of this paper.

Instead, both the *optimal residual vector* $\tilde{e}_{TLS}$ and the optimal residual matrix $(\tilde{E}_A)_{TLS}$ are readily available through (40) and (43) as

$$\tilde{e}_{TLS} = \hat{\lambda} = (y - A\hat{\xi}_{TLS}) \cdot (1 + \hat{\xi}_{TLS}^T \hat{\xi}_{TLS})^{-1}, \tag{50}$$

and through (41) as

$$(\tilde{E}_A)_{TLS} = -\tilde{e}_{TLS} \cdot \hat{\xi}_{TLS}^T = -(y - A\hat{\xi}_{TLS}) \cdot (1 + \hat{\xi}_{TLS}^T \hat{\xi}_{TLS})^{-1} \cdot \hat{\xi}_{TLS}. \tag{51}$$

As optimal variance component estimate, it is now proposed to use the formula

$$(\hat{\sigma}_0^2)_{TLS} = (n - m)^{-1} \cdot [\tilde{e}_{TLS}^T \tilde{e}_{TLS} + (\tilde{e}_A)_{TLS}^T (\tilde{e}_A)_{TLS}] =$$

$$= (n-m)^{-1} \cdot \hat{\lambda}^T \hat{\lambda} (1 + \hat{\xi}_{TLS}^T \hat{\xi}_{TLS}) = \hat{v}_{TLS}/(n-m), \qquad (52)$$

in analogy to the previous estimates (11) and (32).

3.2. A New Model: The EIV-Model with Random Effects (EIV-REM)

In the following, the above EIV-Model (33-34) is *strengthened* by introducing stochastic prior information (p. i.) on the parameters which thereby change their character and become "random effects" as in (13-15). The EIV-REM can, therefore, be stated as

$$\underbrace{y = (A - E_A) \cdot x + e}_{n \times m}, \ q := \text{rk } A \le \min\{m,n\}, \ \beta_0 = x + e_0 \text{ (given)}, \qquad (53)$$

with

$$\begin{bmatrix} e \\ e_A = \text{vec } E_A \\ e_0 \end{bmatrix} \sim \left(\begin{bmatrix} 0 \\ 0 \\ 0 \end{bmatrix}, \sigma_0^2 \begin{bmatrix} I_n & 0 & 0 \\ 0 & I_{mn} & 0 \\ 0 & 0 & Q_0 \end{bmatrix} \right), \ Q_0 \text{ symmetric and nnd.} \qquad (54)$$

The first set of formulas will be derived by assuming that the weight matrix $P_0 := Q_0^{-1}$ exists uniquely for the p. i. Then, the "*TLS collocation (TLSC)*" may be based on the principle

$$e^T e + e_A^T e_A + e_0^T P_0 e_0 = \min \text{ s.t. } (53\text{-}54), \qquad (55)$$

resp. on the *Lagrange function*

$$\Phi(e, e_A, e_0, \lambda) := e^T e + e_A^T e_A + e_0^T P_0 e_0 + 2\lambda^T [(y - A\beta_0 - e + (\beta_0^T \otimes I_n) e_A + A e_0 \underbrace{-E_A e_0}_{=-(e_0^T \otimes I_n) \cdot e_A}] \qquad (56)$$

which needs to be made stationary. The necessary Euler-Lagrange conditions then read:

$$\frac{1}{2} \frac{\partial \Phi}{\partial e} = \tilde{e} - \hat{\lambda} \doteq 0 \ \Rightarrow \ \hat{\lambda} = \tilde{e} \qquad (57)$$

$$\frac{1}{2} \frac{\partial \Phi}{\partial e_A} = \tilde{e}_A + [(\beta_0 - \tilde{e}_0) \otimes I_n] \hat{\lambda} \doteq 0 \ \Rightarrow \ \tilde{E}_A = -\hat{\lambda} \cdot (\beta_0 - \tilde{e}_0)^T =: -\hat{\lambda} \cdot \tilde{x}_{TLSC}^T \qquad (58)$$

$$\frac{1}{2} \frac{\partial \Phi}{\partial e_0} = P_0 \tilde{e}_0 + (A - \tilde{E}_A)^T \hat{\lambda} \doteq 0 \ \Rightarrow \ A^T \hat{\lambda} = \tilde{E}_A^T \hat{\lambda} - P_0 \tilde{e}_0 \ \Rightarrow \qquad (59)$$

$$\Rightarrow A^T \hat{\lambda} = -\tilde{x}_{TLSC} \cdot (\hat{\lambda}^T \hat{\lambda}) + \hat{v}_{TLSC}^0 \text{ for } \hat{v}_{TLSC}^0 := -P_0 \tilde{e}_0 = P_0 (\beta_0 - \tilde{x}_{TLSC}) \qquad (60)$$

$$\frac{1}{2} \frac{\partial \Phi}{\partial \lambda} = y - A(\beta_0 - \tilde{e}_0) - \tilde{e} + \tilde{E}_A (\beta_0 - \tilde{e}_0) \doteq 0 \ \Rightarrow \ y - A \tilde{x}_{TLSC} = \hat{\lambda} (1 + \tilde{x}_{TLSC}^T \tilde{x}_{TLSC}) \ \Rightarrow \qquad (61)$$

$$\Rightarrow \hat{\lambda} = (y - A \tilde{x}_{TLSC}) \cdot (1 + \tilde{x}_{TLSC}^T \tilde{x}_{TLSC})^{-1}. \qquad (62)$$

Combining (60) with (62) results in

$$(c - N \tilde{x}_{TLSC}) \cdot (1 + \tilde{x}_{TLSC}^T \tilde{x}_{TLSC})^{-1} = A^T \hat{\lambda} = -\tilde{x}_{TLSC}(1 + \tilde{x}_{TLSC}^T \tilde{x}_{TLSC})^{-2}(y - A \tilde{x}_{TLSC})^T (y - A \tilde{x}_{TLSC}) + \hat{v}_{TLSC}^0, \qquad (63)$$

and finally in

$$\boxed{(N + (1 + \tilde{x}_{TLSC}^T \tilde{x}_{TLSC}) P_0 - \hat{v}_{TLSC} I_m) \tilde{x}_{TLSC} = c + P_0 \beta_0 \cdot (1 + \tilde{x}_{TLSC}^T \tilde{x}_{TLSC})} \qquad (64)$$

where

$$\hat{v}_{TLSC} := (1 + \tilde{x}_{TLSC}^T \tilde{x}_{TLSC})^{-1}(y - A\tilde{x}_{TLSC})^T(y - A\tilde{x}_{TLSC}), \text{ and}$$
$$\hat{v}^0_{TLSC} := -P_0(\beta_0 - \tilde{x}_{TLSC}) = -P_0\tilde{e}_0, \text{ provided that } P_0 \text{ exists}. \tag{65}$$

In the more general case of a *singular* matrix Q_0, an approach similar to (20) can be followed, leading to the equation system

$$[(1 + \tilde{x}_{TLSC}^T \tilde{x}_{TLSC}) \cdot I_m + Q_0 N - \hat{v}_{TLSC} \cdot Q_0] \tilde{x}_{TLSC} = \beta_0 \cdot (1 + \tilde{x}_{TLSC}^T \tilde{x}_{TLSC}) + Q_0 \underset{\sim}{c} \tag{66}$$

that needs to be solved in connection with (65). Obviously,

$$\begin{array}{ll} \tilde{x}_{TLSC} \longmapsto \beta_0 & \text{if } Q_0 \longmapsto 0, \text{ and} \\ \tilde{x}_{TLSC} \longmapsto \tilde{x}_{TLS} & \text{if } P_0 \longmapsto 0. \end{array} \tag{67}$$

Again, it is still unclear if $\tilde{x}_{TLSC}$ represents an unbiased prediction of the vector x of random effects. Also, very little (if anything) is known about the corresponding MSE-matrix of $\tilde{x}_{TLSC}$. The answers to these open problems will be left for a future contribution. It is, however, possible to find the respective *residual vectors/matrices* represented as follows:

$$\tilde{e}_{TLSC} = \hat{\lambda} = (y - A\tilde{x}_{TLSC}) \cdot (1 + \tilde{x}_{TLSC}^T \tilde{x}_{TLSC})^{-1}, \tag{68}$$

$$(\tilde{E}_A)_{TLSC} = -\hat{\lambda} \cdot \tilde{x}_{TLSC}^T = -(y - A\tilde{x}_{TLSC}) \cdot (1 + \tilde{x}_{TLSC}^T \tilde{x}_{TLSC})^{-1} \cdot \tilde{x}_{TLSC}^T, \tag{69}$$

$$(\tilde{e}_0)_{TLSC} = -Q_0 \cdot \hat{v}^0_{TLSC} = \beta_0 - \tilde{x}_{TLSC}, \tag{70}$$

while a suitable formula for the variance component is suggested as

$$(\hat{\sigma}_0^2)_{TLSC} = n^{-1} \cdot [\hat{v}_{TLSC} + (\hat{v}^0_{TLSC})^T Q_0 (\hat{v}^0_{TLSC})]. \tag{71}$$

4. Conclusions and Outlook

Key formulas have been developed successfully to optimally determine the parameters and residuals within the new 'EIV-Model with p. i.' (or EIV-REM) which turns out to be more general than the other three models considered here (GMM, REM, EIV-Model). In particular, it is quite obvious that

- EIV-REM becomes the REM if $D\{e_A\} := 0$,
- EIV-REM becomes the EIV-Model if $P_0 := 0$,
- EIV-REM becomes the GMM if both $P_0 := 0$ and $D\{e_A\} := 0$.

Hence the new EIV-REM can indeed serve as a universal representative of the whole class of models presented here.

Therefore, in a follow-up paper, it is planned to also cover more general dispersion matrices for e and e_A in (54), similarly to the work by Schaffrin et al. [16] for the EIV-Model with singular dispersion matrices for e_A.

Funding: This research received no external funding.

Conflicts of Interest: The author declares no conflict of interest.

References

1. Rao, C.R. Estimation of parameters in a linear model. *Ann. Statist.* **1976**, *4*, 1023–1037. [CrossRef]
2. Koch, K.-R. *Parameter Estimation and Hypothesis Testing in Linear Models*; Springer: Berlin/Heidelberg, Germany, 1999; p. 334.

3. Moritz, H. A generalized least-squares model. *Studia Geophys. Geodaet.* **1970**, *14*, 353–362. [CrossRef]
4. Schaffrin, B. *Model Choice and Adjustment Techniques in the Presence of Prior Information*; Technical Report No. 351; The Ohio State University Department of Geodetic Science and Survey: Columbus, OH, USA, 1983; p. 37.
5. Golub, G.H.; van Loan, C.F. An analysis of the Total Least-Squares problem. *SIAM J. Numer. Anal.* **1980**, *17*, 883–893. [CrossRef]
6. Van Huffel, S.; Vandewalle, J. *The Total Least-Squares Problem: Computational Aspects and Analysis*; SIAM: Philadelphia, PA, USA, 1991; p. 300.
7. Schaffrin, B.; Wieser, A. On weighted Total Least-Squares adjustment for linear regression. *J. Geod.* **2008**, *82*, 415–421. [CrossRef]
8. Schaffrin, B. Adjusting the Errors-In-Variables model: Linearized Least-Squares vs. nonlinear Total Least-Squares procedures. In *VIII Hotine-Marussi Symposium on Mathematical Geodesy*; Sneeuw, N., Novák, P., Crespi, M., Sansò, F., Eds.; Springer International Publishing: Cham, Switzerland, 2015; pp. 301–307.
9. Schaffrin, B. Total Least-Squares Collocation: The Total Least-Squares approach to EIV-Models with prior information. In Proceedings of the 18th International Workshop on Matrices and Statistics, Smolenice Castle, Slovakia, 23–27 June 2009.
10. Snow, K.; Schaffrin, B. Weighted Total Least-Squares Collocation with geodetic applications. In Proceedings of the SIAM Conference on Applied Linear Algebra, Valencia, Spain, 18–22 June 2012.
11. Schaffrin, B.; Snow, K. Progress towards a rigorous error propagation for Total Least-Squares estimates. *J. Appl. Geod.* **2020**. accepted for publication. [CrossRef]
12. Grafarend, E.; Schaffrin, B. *Adjustment Computations in Linear Models*; Bibliographisches Institut: Mannheim, Germany; Wiesbaden, Germany, 1993; p. 483. (In German)
13. Harville, D.A. Using ordinary least-squares software to compute combined intra-interblock estimates of treatment contrasts. *Am. Stat.* **1986**, *40*, 153–157.
14. Amiri-Simkooei, A.R.; Zangeneh-Nejad, F.; Asgari, J. On the covariance matrix of weighted Total Least-Squares estimates. *J. Surv. Eng.* **2016**, *142*. [CrossRef]
15. Schaffrin, B.; Snow, K. Towards a more rigorous error propagation within the Errors-In-Variables Model for applications in geodetic networks. In Proceedings of the 4th Joint International Symposium on Deformation Monitoring (JISDM 2019), Athens, Greece, 15–17 May 2019. electronic proceedings only.
16. Schaffrin, B.; Snow, K.; Neitzel, F. On the Errors-In-Variables Model with singular dispersion matrices. *J. Geod. Sci.* **2014**, *4*, 28–36. [CrossRef]

© 2020 by the author. Licensee MDPI, Basel, Switzerland. This article is an open access article distributed under the terms and conditions of the Creative Commons Attribution (CC BY) license (http://creativecommons.org/licenses/by/4.0/).

Article

Weighted Total Least Squares (WTLS) Solutions for Straight Line Fitting to 3D Point Data

Georgios Malissiovas [1,*], Frank Neitzel [1], Sven Weisbrich [1] and Svetozar Petrovic [1,2]

1. Institute of Geodesy and Geoinformation Science, Technische Universität Berlin, Strasse des 17. Juni 135, 10623 Berlin, Germany; frank.neitzel@tu-berlin.de (F.N.); s.weisbrich@tu-berlin.de (S.W.); svetozar.petrovic@campus.tu-berlin.de (S.P.)
2. GFZ German Research Centre for Geosciences, Section 1.2: Global Geomonitoring and Gravity Field, Telegrafenberg, 14473 Potsdam, Germany
* Correspondence: georgios.malissiovas@tu-berlin.de

Received: 14 July 2020; Accepted: 25 August 2020; Published: 29 August 2020

Abstract: In this contribution the fitting of a straight line to 3D point data is considered, with Cartesian coordinates x_i, y_i, z_i as observations subject to random errors. A direct solution for the case of equally weighted and uncorrelated coordinate components was already presented almost forty years ago. For more general weighting cases, iterative algorithms, e.g., by means of an iteratively linearized Gauss–Helmert (GH) model, have been proposed in the literature. In this investigation, a new direct solution for the case of pointwise weights is derived. In the terminology of total least squares (TLS), this solution is a direct weighted total least squares (WTLS) approach. For the most general weighting case, considering a full dispersion matrix of the observations that can even be singular to some extent, a new iterative solution based on the ordinary iteration method is developed. The latter is a new iterative WTLS algorithm, since no linearization of the problem by Taylor series is performed at any step. Using a numerical example it is demonstrated how the newly developed WTLS approaches can be applied for 3D straight line fitting considering different weighting cases. The solutions are compared with results from the literature and with those obtained from an iteratively linearized GH model.

Keywords: 3D straight line fitting; total least squares (TLS); weighted total least squares (WTLS); nonlinear least squares adjustment; direct solution; singular dispersion matrix; laser scanning data

---

1. Introduction

Modern geodetic instruments, such as terrestrial laser scanners, provide the user directly with 3D coordinates in a Cartesian coordinate system. However, in most cases these 3D point data are not the final result. For an analysis of the recorded data or for a representation using computer-aided design (CAD), a line, curve or surface approximation with a continuous mathematical function is required.

In this contribution the fitting of a spatial straight line is discussed considering the coordinate components x_i, y_i, z_i of each point P_i as observations subject to random errors, which results in a nonlinear adjustment problem. An elegant direct least squares solution for the case of equally weighted and uncorrelated observations has already been presented in 1982 by Jovičić et al. [1]. Unfortunately, this article was very rarely cited in subsequent publications, which is probably due to the fact that it was written in Croatian language. Similar least squares solutions, direct as well, have been published by Kahn [2] and Drixler ([3], pp. 46–47) some years later. In these contributions, it was shown that the problem of fitting a straight line to 3D point data can be transformed into an eigenvalue problem. An iterative least squares solution for fitting a straight line to equally weighted and uncorrelated 3D points has been presented by Späth [4], by minimizing the sum of squared orthogonal distances of the observed points to the requested straight line.

For more general weighting schemes iterative least squares solutions have been presented by Kupferer [5] and Snow and Schaffrin [6]. In both contributions various nonlinear functional models were introduced and tested and the Gauss-Newton approach has been employed for an iterative least squares solution, which involves the linearization of the functional model to solve the adjustment problem within the Gauss–Markov (GM) or the Gauss–Helmert (GH) models. The linearization of originally nonlinear functional models is a very popular approach in adjustment calculation, where the solution is then determined by iteratively linearized GM or GH models, see e.g., the textbooks by Ghilani [7], Niemeier [8] or Perovic [9]. Pitfalls to be avoided in the iterative adjustment of nonlinear problems have been pointed out by Pope [10] and Lenzmann and Lenzmann [11].

Fitting a straight line to 3D point data can also be considered as an adjustment problem of type total least squares (TLS) for an errors-in-variables (EIV) model, as already pointed out by Snow and Schaffrin [6]. For some weighting cases, problems expressed within the EIV model can have a direct solution using singular value decomposition (SVD). This solution is known under the name Total Least Squares (TLS) and has been firstly proposed by Golub and Van Loan [12], Van Huffel and Vandewalle [13] or Van Huffel and Vandewalle ([14], p. 33 ff.). The TLS solution is obtained by computing the roots of a polynomial, i.e., by solving the characteristic equation of the eigenvalues, which is identical in the case of fitting a straight line to 3D point data to the solutions of Jovičić et al. [1], Kahn [2] and Drixler ([3], pp. 46–47). This is something that has been observed by Malissiovas et al. [15], where a relationship has been presented between TLS and the direct least squares solution of the same adjustment problem, while postulating uncorrelated and equally weighted observations.

To involve more general weight matrices in the adjustment procedure, iterative algorithms have been presented in the TLS literature without linearizing the underlying problem by Taylor series at any step of the solution process. These are algorithmic approaches known as weighted total least squares (WTLS), presented e.g., by Schaffrin and Wieser [16], Shen et al. [17] or Amiri and Jazaeri [18]. A good overview of such algorithms, as well as alternative solution strategies can be found in the dissertation theses of Snow [19] and Malissiovas [20]. An attempt to find a WTLS solution for straight line fitting to 3D point data was made by Guo et al. [21].

To avoid confusion, it is to clarify that the terms TLS and WTLS refer to algorithmic approaches for obtaining a least squares solution, which is either direct or iterative but without linearizing the problem by Taylor series at any step. This follows the statement of Snow ([19], p. 7), that "the terms total least-squares (TLS) and TLS solution [...] will mean the least-squares solution within the EIV model without linearization". Of course, a solution within the GH model is more general in the sense that it can be utilized to solve any nonlinear adjustment problem, while TLS and WTLS algorithms can treat only a certain class of nonlinear adjustment problems. This has been firstly discussed by Neitzel and Petrovic [22] and Neitzel [23], who showed that the TLS estimate within an EIV model can be identified as a special case of the method of least squares within an iteratively linearized GH model.

To the extent of our knowledge, a WTLS algorithm for fitting a straight line to 3D point data has not been presented yet. Therefore, in this study we derive two novel WTLS algorithms for the discussed adjustment problem considering two different cases of stochastic models:

(i) pointwise weights, i.e., coordinate components with same precision for each point and no correlations between them,
(ii) general weights, i.e., correlated coordinate components of individual precision including singular dispersion matrices.

The adjustment problem resulting from case (i) can still be solved directly, i.e., a direct WTLS solution, presented in Section 2.1 of this paper. For case (ii) an iterative solution without linearizing the problem by Taylor series is derived in Section 2.2, i.e., an iterative WTLS solution. Both solutions are based on the work of Malissiovas [20], where similar algorithms have been presented for the solution of other typical geodetic tasks, such as straight line fitting to 2D point data, plane fitting to 3D point data and 2D similarity coordinate transformation. The WTLS solution for straight line fitting to 3D point data will be derived from a geodetic point of view by means of introducing residuals for all observations, formulating appropriate condition and constraint equations, setting up and solving the resulting normal equation systems.

2. Straight Line Fitting to 3D Point Data

A straight line in 3D space can be expressed by

$$\frac{y - y_0}{a} = \frac{x - x_0}{b} = \frac{z - z_0}{c}, \quad (1)$$

as explained e.g., in the handbook of mathematics by Bronhstein et al. ([24], p. 217). This equation defines a line that passes through a point with coordinates x_0, y_0 and z_0 and is parallel to a direction vector with components a, b and c. Since the number of unknown parameters is six, two additional constraints between the unknown parameters have to be taken into account for a solution, as Snow and Schaffrin [6] have clearly explained. Proper constraints will be selected at a later point of the derivations.

Considering an overdetermined configuration for which we want to obtain a least squares solution, we introduce for all points P_i residuals $v_{x_i}, v_{y_i}, v_{z_i}$ for the observations x_i, y_i, z_i assuming that they are normally distributed with $\mathbf{v} \sim (\mathbf{0}, \sigma_0^2 \, \mathbf{Q}_{\text{LL}})$. Here the $1 \times 3n$ vector $\mathbf{v}$ contains the residuals $v_{x_i}, v_{y_i}, v_{z_i}$, $\mathbf{Q}_{\text{LL}}$ is the $3n \times 3n$ corresponding cofactor matrix and σ_0^2 the theoretical variance factor. Based on (1), the nonlinear conditions equations

$$a(x_i + v_{x_i} - x_0) - b(y_i + v_{y_i} - y_0) = 0,$$
$$b(z_i + v_{z_i} - z_0) - c(x_i + v_{x_i} - x_0) = 0, \quad (2)$$
$$c(y_i + v_{y_i} - y_0) - a(z_i + v_{z_i} - z_0) = 0,$$

can be formulated for each point P_i, with $i = 1, \ldots, n$ and n being the number of observed points. A functional model for this problem can be expressed by two of these three nonlinear condition equations per observed point. Using all three condition equations for solving the problem would lead to a singular normal matrix due to linearly dependent normal equations.

2.1. Direct Total Least Squares Solution for Equally Weighted and Uncorrelated Observations

Considering all coordinate components x_i, y_i, z_i for all points P_i as equally weighted and uncorrelated observations with

$$p_{y_i} = p_{x_i} = p_{z_i} = 1 \; \forall i, \quad (3)$$

a least squares solution can be derived by minimizing the objective function

$$\Omega = \sum_{i=1}^{n} v_{x_i}^2 + v_{y_i}^2 + v_{z_i}^2 \to \min. \quad (4)$$

A direct least squares solution of this problem, respectively a TLS solution, has been presented by Malissiovas et al. [15] and Malissiovas [20]. According to these investigations, equally weighted residuals correspond to the normal distances

$$D_i^2 = v_{x_i}^2 + v_{y_i}^2 + v_{z_i}^2, \quad (5)$$

as deviation measures between the observed 3D point cloud and the straight line to be computed. Thus, the objective function can be written equivalently as

$$\Omega = \sum_{i=1}^{n} v_{x_i}^2 + v_{y_i}^2 + v_{z_i}^2 = \sum_{i=1}^{n} D_i^2 \to \min. \tag{6}$$

An expression for the squared normal distances

$$D^2 = \frac{[a(x-x_0) - b(y-y_0)]^2 + [b(z-z_0) - c(x-x_0)]^2 + [c(y-y_0) - a(z-z_0)]^2}{a^2 + b^2 + c^2}, \tag{7}$$

can be found in the handbook of Bronhstein et al. ([24], p. 218).

An appropriate parameterization of the problem involves the substitution of the unknown parameters y_0, x_0 and z_0 with the coordinates of the center of mass of the observed 3D point data. A proof for this parameter replacement has been given by Jovičić et al. [1] and concerns only the case of equally weighted and uncorrelated observations. A solution for the line parameters can be computed by minimizing the objective function (6), under the constraint

$$a^2 + b^2 + c^2 = 1, \tag{8}$$

or by searching for stationary points of the Lagrange function

$$K = \sum_{i=1}^{n} D_i^2 - k(a^2 + b^2 + c^2 - 1), \tag{9}$$

with k denoting the Lagrange multiplier. The line parameters can be computed directly from the normal equations, either by solving a characteristic cubic equation or an eigenvalue problem. A detailed explanation of this approach was given by Malissiovas ([20], p. 74 ff.). In the following subsections we will give up the restriction of equally weighted and uncorrelated coordinate components and derive least squares solutions considering more general weighting schemes for the observations.

2.2. Direct Weighted Total Least Squares Solution

In this case we consider the coordinate components x_i, y_i, z_i of each point P_i to be uncorrelated and of equal precision with

$$\sigma_{y_i} = \sigma_{x_i} = \sigma_{z_i} = \sigma_i \; \forall i, \tag{10}$$

yielding the corresponding (pointwise) weights

$$p_i = \frac{1}{\sigma_i^2} \; \forall i. \tag{11}$$

A least squares solution of the problem can be found by minimizing the objective function

$$\begin{aligned}\Omega &= \sum_{i=1}^{n} p_{x_i} v_{x_i}^2 + p_{y_i} v_{y_i}^2 + p_{z_i} v_{z_i}^2 \to \min \\ &= \sum_{i=1}^{n} p_i \left(v_{x_i}^2 + v_{y_i}^2 + v_{z_i}^2 \right) \to \min.\end{aligned} \tag{12}$$

Taking into account the relation of the squared residuals with the normal distances of Equation (5), it is possible to express the objective function as

$$\Omega = \sum_{i=1}^{n} p_i D_i^2 \to \min. \tag{13}$$

Using the constraint (8), the expression of the normal distances can be written as

$$D_i^2 = [a(x_i - x_0) - b(y_i - y_0)]^2 + [b(z_i - z_0) - c(x_i - x_0)]^2 + [c(y_i - y_0) - a(z_i - z_0)]^2. \tag{14}$$

A further simplification of the problem is possible, by replacing the unknown parameters y_0, x_0 and z_0 with the coordinates of the weighted center of mass of the 3D point data

$$y_0 = \frac{\sum_{i=1}^{n} p_i y_i}{\sum_{i=1}^{n} p_i}, \quad x_0 = \frac{\sum_{i=1}^{n} p_i x_i}{\sum_{i=1}^{n} p_i}, \quad z_0 = \frac{\sum_{i=1}^{n} p_i z_i}{\sum_{i=1}^{n} p_i}. \tag{15}$$

It can be proven that this parameterization is allowed and possible, following the same line of thinking as in the study of Jovičić et al. [1]. Therefore, for this weighted case we can write the normal distances as

$$D_i^2 = (ax_i' - by_i')^2 + (bz_i' - cx_i')^2 + (cy_i' - az_i')^2, \tag{16}$$

with the coordinates reduced to the weighted center of mass of the observed 3D point data

$$y_i' = y_i - \frac{\sum_{i=1}^{n} p_i y_i}{\sum_{i=1}^{n} p_i}, \quad x_i' = x_i - \frac{\sum_{i=1}^{n} p_i x_i}{\sum_{i=1}^{n} p_i}, \quad z_i' = z_i - \frac{\sum_{i=1}^{n} p_i z_i}{\sum_{i=1}^{n} p_i}. \tag{17}$$

Searching for stationary points of the Lagrange function

$$\begin{aligned} K &= \sum_{i=1}^{n} D_i^2 - k(a^2 + b^2 + c^2 - 1) \\ &= \sum_{i=1}^{n} (ax_i' - by_i')^2 + (bz_i' - cx_i')^2 + (cy_i' - az_i')^2 - k(a^2 + b^2 + c^2 - 1) \end{aligned} \tag{18}$$

leads to the system of normal equations

$$\frac{\partial K}{\partial a} = 2a \left(\sum_{i=1}^{n} p_i x_i'^2 + \sum_{i=1}^{n} p_i z_i'^2 - k \right) - 2b \sum_{i=1}^{n} p_i y_i' x_i' - 2c \sum_{i=1}^{n} p_i y_i' z_i' = 0, \tag{19}$$

$$\frac{\partial K}{\partial b} = 2b \left(\sum_{i=1}^{n} p_i y_i'^2 + \sum_{i=1}^{n} p_i z_i'^2 - k \right) - 2a \sum_{i=1}^{n} p_i y_i' x_i' - 2c \sum_{i=1}^{n} p_i x_i' z_i' = 0, \tag{20}$$

$$\frac{\partial K}{\partial c} = 2c \left(\sum_{i=1}^{n} p_i y_i'^2 + \sum_{i=1}^{n} p_i x_i'^2 - k \right) - 2a \sum_{i=1}^{n} p_i y_i' z_i' - 2b \sum_{i=1}^{n} p_i x_i' z_i' = 0, \tag{21}$$

and

$$\frac{\partial K}{\partial k} = -\left(a^2 + b^2 + c^2 - 1 \right) = 0. \tag{22}$$

Equations (19) to (21) are a homogeneous system of equations which are linear in the unknown line parameters a, b and c, and has a nontrivial solution if

$$\begin{vmatrix} (d_1 - k) & d_4 & d_5 \\ d_4 & (d_2 - k) & d_6 \\ d_5 & d_6 & (d_3 - k) \end{vmatrix} = 0, \tag{23}$$

with the elements of the determinant

$$d_1 = \sum_{i=1}^{n} p_i x_i'^2 + \sum_{i=1}^{n} p_i z_i'^2, \quad d_2 = \sum_{i=1}^{n} p_i y_i'^2 + \sum_{i=1}^{n} p_i z_i'^2, \quad d_3 = \sum_{i=1}^{n} p_i y_i'^2 + \sum_{i=1}^{n} p_i x_i'^2,$$
$$d_4 = -\sum_{i=1}^{n} p_i y_i' x_i', \quad d_5 = -\sum_{i=1}^{n} p_i y_i' z_i', \quad d_6 = -\sum_{i=1}^{n} p_i x_i' z_i'. \tag{24}$$

Equation (23) is a cubic characteristic equation with the unknown parameter k and a solution is given by Bronhstein et al. ([24], p. 63). The unknown line parameters a, b and c can be computed either by substituting k_{min} into Equations (19)–(21) under the constraint (22) or by transforming the equation system and solving an eigenvalue problem, i.e., the direct WTLS solution.

2.3. Iterative Weighted Total Least Squares Solution

In this case, we consider the fact that the coordinate components x_i, y_i, z_i of each point P_i are not original observations. They are either

(i) derived from single point determinations, e.g., using polar elementary measurements of slope distance ρ_i, direction ϕ_i and tilt angle θ_i, e.g., from measurements with a terrestrial laser scanner, so that the coordinates result from the functional relationship

$$\begin{aligned} x_i &= \rho_i \cos \phi_i \sin \theta_i, \\ y_i &= \rho_i \sin \phi_i \cos \theta_i, \\ z_i &= \rho_i \cos \theta_i. \end{aligned} \tag{25}$$

Using the standard deviations σ_{ρ_i}, σ_{ϕ_i} and σ_{θ_i} of the elementary measurements, a 3×3 variance-covariance matrix can be provided for the coordinate components x_i, y_i, z_i of each point P_i by variance-covariance propagation based on (25). Covariances among the n different points do not occur in this case.

or

(ii) derived from a least squares adjustment of an overdetermined geodetic network. From the solution within the GM or GH model, the $3n \times 3n$ variance-covariance matrix of the coordinate components x_i, y_i, z_i of all points P_i can be obtained. This matrix is a full matrix, considering also covariances among the n different points. It may even have a rank deficiency in case the coordinates were determined by a free network adjustment.

The variances and covariances from (i) or (ii) can then be assembled in a dispersion matrix

$$\mathbf{Q}_{LL} = \begin{bmatrix} Q_{xx} & Q_{xy} & Q_{xz} \\ Q_{yx} & Q_{yy} & Q_{yz} \\ Q_{zx} & Q_{zy} & Q_{zz} \end{bmatrix}, \tag{26}$$

for the coordinate components as "derived observations". In case of a non-singular dispersion matrix, it is possible to compute the respective weight matrix

$$\mathbf{P} = \begin{bmatrix} P_{xx} & P_{xy} & P_{xz} \\ P_{yx} & P_{yy} & P_{yz} \\ P_{zx} & P_{zy} & P_{zz} \end{bmatrix} = \mathbf{Q}_{LL}^{-1}. \tag{27}$$

Starting with the functional model (2), we choose to work with the nonlinear condition equations

$$\begin{aligned} c(y_i + v_{y_i} - y_0) - a(z_i + v_{z_i} - z_0) &= 0, \\ c(x_i + v_{x_i} - x_0) - b(z_i + v_{z_i} - z_0) &= 0. \end{aligned} \tag{28}$$

Two additional constraints must be taken into account in this case. For example, Snow and Schaffrin [6] proposed the constraints

$$a^2 + b^2 + c^2 = 1,$$
$$ay_0 + bx_0 + cz_0 = 0. \tag{29}$$

A further development of an algorithmic solution using these constraints is possible. However, we choose to parameterize the functional model so that an additional constraint is unnecessary and the equations become simpler. Therefore, we choose to fix one coordinate of the point on the line

$$z_0 = \frac{\sum_{i=1}^{n} z_i}{n} = \bar{z}, \tag{30}$$

as well as one of the unknown components of the direction vector

$$c = 1. \tag{31}$$

This simplification of the functional model has been used by Borovička et al. [25] for a similar practical example, this of estimating the trajectory of a meteor. As already mentioned in that work, the resulting direction vector of the straight line can be transformed to a unit vector afterwards, i.e., a solution using the constraint $a^2 + b^2 + c^2 = 1$. The simplified functional model can be written in vector notation as

$$(\mathbf{y}_c + \mathbf{v}_y - \mathbf{e}y_0) - a(\mathbf{z}_c + \mathbf{v}_z - \mathbf{e}\bar{z}) = 0,$$
$$(\mathbf{x}_c + \mathbf{v}_x - \mathbf{e}x_0) - b(\mathbf{z}_c + \mathbf{v}_z - \mathbf{e}\bar{z}) = 0, \tag{32}$$

with the vectors $\mathbf{x}_c$, $\mathbf{y}_c$ and $\mathbf{z}_c$ containing the coordinates of the observed points and vectors $\mathbf{v}_x$, $\mathbf{v}_y$ and $\mathbf{v}_z$ the respective residuals. Vector $\mathbf{e}$ is a vector of ones, with length equal to the number of observed points n.

A weighted least squares solution of this problem can be derived by minimizing the objective function

$$\Omega = \mathbf{v}_x^T \mathbf{P}_{xx} \mathbf{v}_x + \mathbf{v}_y^T \mathbf{P}_{yy} \mathbf{v}_y + \mathbf{v}_z^T \mathbf{P}_{zz} \mathbf{v}_z + 2\,\mathbf{v}_x^T \mathbf{P}_{xy} \mathbf{v}_y + 2\,\mathbf{v}_x^T \mathbf{P}_{xz} \mathbf{v}_z + 2\,\mathbf{v}_y^T \mathbf{P}_{yz} \mathbf{v}_z \tag{33}$$

or by searching for stationary points of the Lagrange function

$$K = \Omega - 2\mathbf{k}_1^T \left[(\mathbf{y}_c + \mathbf{v}_y - \mathbf{e}y_0) - a(\mathbf{z}_c + \mathbf{v}_z - \mathbf{e}\bar{z})\right]$$
$$- 2\mathbf{k}_2^T \left[(\mathbf{x}_c + \mathbf{v}_x - \mathbf{e}x_0) - b(\mathbf{z}_c + \mathbf{v}_z - \mathbf{e}\bar{z})\right] \tag{34}$$

with two distinct vectors $\mathbf{k}_1$ and $\mathbf{k}_2$ of Lagrange multipliers. Taking the partial derivatives with respect to all unknowns and setting them to zero results in the system of normal equations

$$\frac{1}{2}\frac{\partial K}{\partial \mathbf{v}_x^T} = \mathbf{P}_{xx}\mathbf{v}_x + \mathbf{P}_{xy}\mathbf{v}_y + \mathbf{P}_{xz}\mathbf{v}_z - \mathbf{k}_2 = 0, \tag{35}$$

$$\frac{1}{2}\frac{\partial K}{\partial \mathbf{v}_y^T} = \mathbf{P}_{yy}\mathbf{v}_y + \mathbf{P}_{yx}\mathbf{v}_x + \mathbf{P}_{yz}\mathbf{v}_z - \mathbf{k}_1 = 0, \tag{36}$$

$$\frac{1}{2}\frac{\partial K}{\partial \mathbf{v}_z^T} = \mathbf{P}_{yy}\mathbf{v}_y + \mathbf{P}_{xz}\mathbf{v}_x + \mathbf{P}_{yz}\mathbf{v}_y + a\mathbf{k}_1 + b\mathbf{k}_2 = 0, \tag{37}$$

$$\frac{1}{2}\frac{\partial K}{\partial \mathbf{k}_1^T} = (\mathbf{y}_c + \mathbf{v}_y - \mathbf{e}y_0) - a(\mathbf{z}_c + \mathbf{v}_z - \mathbf{e}\bar{z}) = \mathbf{0}, \tag{38}$$

$$\frac{1}{2}\frac{\partial K}{\partial \mathbf{k}_2^T} = (\mathbf{x}_c + \mathbf{v}_x - \mathbf{e}x_0) - a(\mathbf{z}_c + \mathbf{v}_z - \mathbf{e}\bar{z}) = \mathbf{0}, \tag{39}$$

$$\frac{1}{2}\frac{\partial K}{\partial a} = \mathbf{k}_1^T(\mathbf{z}_c + \mathbf{v}_z - \mathbf{e}\bar{z}) = 0, \tag{40}$$

$$\frac{1}{2}\frac{\partial K}{\partial b} = \mathbf{k}_2^T(\mathbf{z}_c + \mathbf{v}_z - \mathbf{e}\bar{z}) = 0, \tag{41}$$

$$\frac{1}{2}\frac{\partial K}{\partial y_0} = \mathbf{e}^T \mathbf{k}_1 = 0, \tag{42}$$

$$\frac{1}{2}\frac{\partial K}{\partial x_0} = \mathbf{e}^T \mathbf{k}_2 = 0. \tag{43}$$

To derive explicit expressions for the residual vectors, we write Equations (35)–(37) as

$$\begin{bmatrix} \mathbf{P}_{xx} & \mathbf{P}_{xy} & \mathbf{P}_{xz} \\ \mathbf{P}_{yx} & \mathbf{P}_{yy} & \mathbf{P}_{yz} \\ \mathbf{P}_{zx} & \mathbf{P}_{zy} & \mathbf{P}_{zz} \end{bmatrix} \begin{bmatrix} \mathbf{v}_x \\ \mathbf{v}_y \\ \mathbf{v}_z \end{bmatrix} = \begin{bmatrix} \mathbf{k}_2 \\ \mathbf{k}_1 \\ -a\mathbf{k}_1 - b\mathbf{k}_2 \end{bmatrix}, \tag{44}$$

which leads to the solution for the residual vectors

$$\begin{bmatrix} \mathbf{v}_x \\ \mathbf{v}_y \\ \mathbf{v}_z \end{bmatrix} = \begin{bmatrix} \mathbf{P}_{xx} & \mathbf{P}_{xy} & \mathbf{P}_{xz} \\ \mathbf{P}_{yx} & \mathbf{P}_{yy} & \mathbf{P}_{yz} \\ \mathbf{P}_{zx} & \mathbf{P}_{zy} & \mathbf{P}_{zz} \end{bmatrix}^{-1} \begin{bmatrix} \mathbf{k}_2 \\ \mathbf{k}_1 \\ -a\mathbf{k}_1 - b\mathbf{k}_2 \end{bmatrix}$$
$$= \begin{bmatrix} \mathbf{Q}_{xx} & \mathbf{Q}_{xy} & \mathbf{Q}_{xz} \\ \mathbf{Q}_{yx} & \mathbf{Q}_{yy} & \mathbf{Q}_{yz} \\ \mathbf{Q}_{zx} & \mathbf{Q}_{zy} & \mathbf{Q}_{zz} \end{bmatrix} \begin{bmatrix} \mathbf{k}_2 \\ \mathbf{k}_1 \\ -a\mathbf{k}_1 - b\mathbf{k}_2 \end{bmatrix}, \tag{45}$$

or explicitly

$$\mathbf{v}_x = (\mathbf{Q}_{xy} - a\mathbf{Q}_{xz})\mathbf{k}_1 + (\mathbf{Q}_{xx} - b\mathbf{Q}_{xz})\mathbf{k}_2, \tag{46}$$

$$\mathbf{v}_y = (\mathbf{Q}_{yy} - a\mathbf{Q}_{yz})\mathbf{k}_1 + (\mathbf{Q}_{xy} - b\mathbf{Q}_{yz})\mathbf{k}_2, \tag{47}$$

$$\mathbf{v}_z = (\mathbf{Q}_{yz} - a\mathbf{Q}_z)\mathbf{k}_1 + (\mathbf{Q}_{xz} - b\mathbf{Q}_{zz})\mathbf{k}_2. \tag{48}$$

Inserting these expressions for the residual vectors into Equations (38) and (39) yields

$$\mathbf{W}_1 \mathbf{k}_1 + \mathbf{W}_2 \mathbf{k}_2 + \mathbf{y}_c - \mathbf{e}y_0 - a\mathbf{z}_c = \mathbf{0}, \tag{49}$$

$$\mathbf{W}_2 \mathbf{k}_1 + \mathbf{W}_3 \mathbf{k}_2 + \mathbf{x}_c - \mathbf{e}x_0 - b\mathbf{z}_c = \mathbf{0}, \tag{50}$$

with the auxiliary matrices $\mathbf{W}_1$, $\mathbf{W}_2$ and $\mathbf{W}_3$ as

$$\mathbf{W}_1 = \mathbf{Q}_y - 2a\mathbf{Q}_{yz} + a^2\mathbf{Q}_z, \tag{51}$$

$$\mathbf{W}_2 = \mathbf{Q}_{xy} - a\mathbf{Q}_{xz} - b\mathbf{Q}_{yz} + ab\mathbf{Q}_z, \tag{52}$$

$$\mathbf{W}_3 = \mathbf{Q}_x - 2b\mathbf{Q}_{xz} + b^2\mathbf{Q}_z, \tag{53}$$

Using Equations (49), (50) and (40)–(43), the nonlinear equation system

$$\begin{aligned}
\mathbf{W}_1\mathbf{k}_1 + \mathbf{W}_2\mathbf{k}_2 - \mathbf{e}y_0 - a\mathbf{z}_c &= -\mathbf{y}_c, \\
\mathbf{W}_2\mathbf{k}_1 + \mathbf{W}_3\mathbf{k}_2 - \mathbf{e}x_0 - b\mathbf{z}_c &= -\mathbf{x}_c, \\
-(\mathbf{z}_c + \mathbf{v}_z - \mathbf{e}\bar{z})^T \mathbf{k}_1 &= 0, \\
-(\mathbf{z}_c + \mathbf{v}_z - \mathbf{e}\bar{z})^T \mathbf{k}_2 &= 0, \\
-\mathbf{e}^T\mathbf{k}_1 &= 0, \\
-\mathbf{e}^T\mathbf{k}_2 &= 0
\end{aligned} \tag{54}$$

can be set up. To express this system of equations into a block matrix form and to obtain a symmetrical matrix of normal equations in the following, it is advantageous to add the terms $-a\left(\mathbf{v}_z - \mathbf{e}\bar{z}\right)$ and $-b\left(\mathbf{v}_z - \mathbf{e}\bar{z}\right)$ to both sides of the first two equations, leading to the system of equations

$$\begin{aligned}
\mathbf{W}_1\mathbf{k}_1 + \mathbf{W}_2\mathbf{k}_2 - \mathbf{e}y_0 - a\left(\mathbf{z}_c + \mathbf{v}_z - \mathbf{e}\bar{z}\right) &= -\mathbf{y}_c - a\left(\mathbf{v}_z - \mathbf{e}\bar{z}\right), \\
\mathbf{W}_2\mathbf{k}_1 + \mathbf{W}_3\mathbf{k}_2 - \mathbf{e}x_0 - b\left(\mathbf{z}_c + \mathbf{v}_z - \mathbf{e}\bar{z}\right) &= -\mathbf{x}_c - b\left(\mathbf{v}_z - \mathbf{e}\bar{z}\right), \\
-(\mathbf{z}_c + \mathbf{v}_z - \mathbf{e}\bar{z})^T \mathbf{k}_1 &= 0, \\
-(\mathbf{z}_c + \mathbf{v}_z - \mathbf{e}\bar{z})^T \mathbf{k}_2 &= 0, \\
-\mathbf{e}^T\mathbf{k}_1 &= 0, \\
-\mathbf{e}^T\mathbf{k}_2 &= 0.
\end{aligned} \tag{55}$$

Arranging the unknowns in a vector

$$\mathbf{X} = \begin{bmatrix} \mathbf{k}_1 \\ \mathbf{k}_2 \\ a \\ b \\ y_0 \\ x_0 \end{bmatrix}, \tag{56}$$

the equation system (55) can be written as

$$\mathbf{N}\mathbf{X} = \mathbf{n}. \tag{57}$$

with the matrix of normal equations

$$\mathbf{N} = \begin{bmatrix} \mathbf{W} & \mathbf{A} \\ \mathbf{A}^T & \mathbf{0} \end{bmatrix}, \tag{58}$$

constructed using the matrices

$$\mathbf{W} = \begin{bmatrix} \mathbf{W}_1 & \mathbf{W}_2 \\ \mathbf{W}_2 & \mathbf{W}_3 \end{bmatrix} \tag{59}$$

and

$$\mathbf{A}^T = \begin{bmatrix} -(\mathbf{z}_c + \mathbf{v}_z - e\bar{z}) & 0 \\ 0 & -(\mathbf{z}_c + \mathbf{v}_z - e\bar{z}) \\ -\mathbf{e} & 0 \\ 0 & -\mathbf{e} \end{bmatrix}. \tag{60}$$

The right-hand side of the equation system (57) reads

$$\mathbf{n} = \begin{bmatrix} -\mathbf{y}_c - a\,(\mathbf{z}_c + \mathbf{v}_z - e\bar{z}) \\ -\mathbf{x}_c - b\,(\mathbf{z}_c + \mathbf{v}_z - e\bar{z}) \\ 0 \\ 0 \\ 0 \\ 0 \end{bmatrix} \tag{61}$$

It is important to point out that matrix $\mathbf{N}$ and vector $\mathbf{n}$ contain the unknown parameters a, b and $\mathbf{v}_z$ in their entries. Therefore, to express and solve these normal equations that live in the "nonlinear world" with the help of vectors and matrices (that only exist in the "linear world"), appropriate approximate values a^0, b^0 and $\mathbf{v}_z^0$ have to be introduced for all auxiliary matrices required for the setup of matrix $\mathbf{N}$ and vector $\mathbf{n}$. The solution for the vector of unknowns can be computed by

$$\hat{\mathbf{X}} = \mathbf{N}^{-1}\mathbf{n}, \tag{62}$$

without the need of a linearization by Taylor series at any step of the calculation process. The WTLS solution for the line parameters can be computed following the *ordinary iteration method*, as it is explained for example by Bronhstein ([24], p. 896). Therefore, the solutions $\hat{a}$, $\hat{b}$ and $\tilde{\mathbf{v}}_z$, after stripping them of their random character, are to be substituted as new approximate values as long as necessary until a sensibly chosen break-off condition is met. For example, the maximum absolute difference between two consecutive solutions must be smaller than a predefined threshold ϵ, which in this problem can be formulated as

$$\max \left| \begin{bmatrix} a^0 \\ b^0 \end{bmatrix} - \begin{bmatrix} \hat{a} \\ \hat{b} \end{bmatrix} \right| \leq \epsilon, \tag{63}$$

with $|\cdot|$ denoting the absolute value. The predicted residual vectors $\tilde{\mathbf{v}}_x$, $\tilde{\mathbf{v}}_y$, $\tilde{\mathbf{v}}_z$ can be computed from Equations (46)–(48). The iterative procedure for the presented WTLS solution can be found in Algorithm 1.

Algorithm 1 Iterative WTLS solution
―――
Choose approximate values for a^0, b^0 and $\mathbf{v}_z^0$.
Define parameter $c = 1$.
Set threshold ϵ for the break-off condition of the iteration process.
Set parameter $d_a = d_b = \infty$, for entering the iteration process.
while $d_a > \epsilon$ and $d_b > \epsilon$ **do**
    Compute matrices $\mathbf{W}$ and $\mathbf{A}$.
    Build matrix $\mathbf{N}$ and vector $\mathbf{n}$.
    Estimate the vector of unknowns $\hat{\mathbf{X}}$.
    Compute the residual vector $\tilde{\mathbf{v}}_z$.
    Compute parameters $d_a = |\hat{a} - a^0|$ and $d_b = |\hat{b} - b^0|$.
    Update the approximate values with the estimated ones, with $a^0 = \hat{a}$, $b^0 = \hat{b}$ and $\mathbf{v}_z^0 = \tilde{\mathbf{v}}_z$.
end while
return $\hat{a}$ and $\hat{b}$, with $c = 1$.
―――

After computing the line parameters $\hat{a}$ and $\hat{b}$ and putting them into a vector, we can easily scale it into a unit vector by dividing each component with the length of the vector

$$\begin{bmatrix} a_n \\ b_n \\ c_n \end{bmatrix} = \frac{\begin{bmatrix} \hat{a} \\ \hat{b} \\ 1 \end{bmatrix}}{\left\| \begin{bmatrix} \hat{a} \\ \hat{b} \\ 1 \end{bmatrix} \right\|} \tag{64}$$

with $||\cdot||$ denoting the Euclidean norm. The derived parameters a_n, b_n and c_n refer to the normalized components of a unit vector, that is parallel to the requested line, with

$$a_n^2 + b_n^2 + c_n^2 = 1 \tag{65}$$

2.4. WTLS Solution with Singular Dispersion Matrices

The algorithmic approach presented in Section 2.3 can also cover cases when dispersion matrices are singular. Such a solution depends on the inversion of matrix $\mathbf{N}$ (58), which depends on the rank deficiency of matrix $\mathbf{W}$ (59). Following the argumentation of Malissiovas ([20], p. 112), a criterion that ensures a unique solution of the problem can be described in this case by

$$\text{rank}\left(\left[\mathbf{W} \mid \mathbf{A}\right]\right) = 2n, \tag{66}$$

with

- rank of $\mathbf{W} \leq 2n$, with $2n$ = number of condition equations, since for each of the observed n points two condition equations from Equation (2) are taken into account;
- rank of $\mathbf{A} = m$, with m = number of unknown parameters.

In cases of singular dispersion matrices, the rank of matrix $\mathbf{W}$ will be smaller than $2n$. A unique solution will exist if the rank of the augmented matrix $[\mathbf{W} \mid \mathbf{A}]$ is equal to the number of condition equations $2n$ of the problem. It is important to mention that the developed idea is based on the Neitzel–Schaffrin criterion, which has been firstly proposed by Neitzel and Schaffrin [26,27], particularly for a solution of an adjustment problem within the GH model when singular dispersion matrices must be employed.

2.5. A Posteriori Error Estimation

In this section we want to determine the variance-covariance matrix of the estimated parameters. The following derivations can be used for computing the a posteriori stochastic information for all weighted cases discussed in this investigation, i.e., the direct and the iterative WTLS solutions. Therefore, we will employ the fundamental idea of variance-covariance propagation. This is a standard procedure explained by many authors, like for example in the textbooks of Wells and Krakiwsky ([28], p. 20), Mikhail ([29], p. 76 ff.) or Niemeier ([8], p. 51 ff.) and has been further employed in the GM and the GH model, so that the a posteriori stochastic results can be computed using directly the developed matrices from each model. A detailed explanation is given by Malissiovas ([20], p. 31 ff.).

As we have already mentioned in the introduction of this article, a TLS, respectively a WTLS solution, can be regarded as a special case of a least squares solution within the GH model. From the presented WTLS algorithm we observe that the derived matrix of normal Equation (58) is equal to the matrix if it was computed within the GH model. Therefore, it is possible to compute

$$\mathbf{N}^{-1} = \begin{bmatrix} \mathbf{Q}_{11} & \mathbf{Q}_{12} \\ \mathbf{Q}_{21} & \mathbf{Q}_{22} \end{bmatrix} \tag{67}$$

and extract the dispersion matrix for the unknown parameters from

$$\mathbf{Q}_{\hat{x}\hat{x}} = -\mathbf{Q}_{22}. \tag{68}$$

The a posteriori variance factor is

$$\hat{\sigma}_0^2 = \frac{\mathbf{v}^\mathsf{T} \mathbf{P} \mathbf{v}}{r} \tag{69}$$

with vector $\mathbf{v}$ holding all residuals and r denoting the redundancy of the adjustment problem. In case of a singular dispersion matrix, it is not possible to compute the weight matrix $\mathbf{P}$, as in Equation (27). Therefore, we can make use of the solution for the residual vectors from Equation (45) and insert them in Equation (69) to obtain

$$\hat{\sigma}_0^2 = \frac{\mathbf{v}_x^\mathsf{T} \mathbf{k}_2 + \mathbf{v}_y^\mathsf{T} \mathbf{k}_1 - a\mathbf{v}_z^\mathsf{T} \mathbf{k}_1 - b\mathbf{v}_z^\mathsf{T} \mathbf{k}_2}{r}. \tag{70}$$

The variance-covariance matrix of the unknown parameters can be then derived by

$$\Sigma_{\hat{x}\hat{x}} = \hat{\sigma}_0^2 \mathbf{Q}_{\hat{x}\hat{x}} = \begin{bmatrix} \sigma_a^2 & \sigma_{ab} & \sigma_{ay_0} & \sigma_{ax_0} \\ \sigma_{ba} & \sigma_b^2 & \sigma_{by_0} & \sigma_{bx_0} \\ \sigma_{y_0 a} & \sigma_{y_0 b} & \sigma_{y_0}^2 & \sigma_{y_0 x_0} \\ \sigma_{x_0 a} & \sigma_{x_0 b} & \sigma_{x_0 y_0} & \sigma_{x_0}^2 \end{bmatrix}. \tag{71}$$

To derive the variance-covariance matrix of the normalized vector components (64), we can explicitly write the equations

$$a_n = \frac{\hat{a}}{\sqrt{\hat{a}^2 + \hat{b}^2 + c^2}},$$

$$b_n = \frac{\hat{b}}{\sqrt{\hat{a}^2 + \hat{b}^2 + c^2}}, \tag{72}$$

$$c_n = \frac{c}{\sqrt{\hat{a}^2 + \hat{b}^2 + c^2}}.$$

with $c = 1$. Following the standard procedure of the variance-covariance propagation in nonlinear cases, we can write the Jacobian matrix

$$\mathbf{F} = \begin{bmatrix} \dfrac{\partial a_n}{\partial a} & \dfrac{\partial a_n}{\partial b} \\ \dfrac{\partial b_n}{\partial a} & \dfrac{\partial b_n}{\partial b} \\ \dfrac{\partial c_n}{\partial a} & \dfrac{\partial c_n}{\partial b} \end{bmatrix}. \qquad (73)$$

Taking into account the variances and covariances of the line parameters $\hat{a}$ and $\hat{b}$ from (71)

$$\Sigma_{\hat{a}\hat{b}} = \Sigma_{\hat{x}\hat{x}}(1:2,1:2) = \begin{bmatrix} \sigma_a^2 & \sigma_{ab} \\ \sigma_{ba} & \sigma_b^2 \end{bmatrix}, \qquad (74)$$

we can compute the variance-covariance matrix of the normalized components

$$\Sigma_{a_n b_n c_n} = \mathbf{F} \Sigma_{\hat{a}\hat{b}} \mathbf{F}^T = \begin{bmatrix} \sigma_{a_n}^2 & \sigma_{a_n b_n} & \sigma_{a_n c_n} \\ \sigma_{b_n a_n} & \sigma_{b_n}^2 & \sigma_{b_n c_n} \\ \sigma_{c_n a_n} & \sigma_{c_n b_n} & \sigma_{c_n}^2 \end{bmatrix}. \qquad (75)$$

3. Numerical Examples

In this section we present the solutions for fitting a straight line to 3D point data using the TLS approach from Section 2.1 and the newly developed WTLS approaches from Sections 2.2 and 2.3. The utilized dataset for this investigaiton consists of $n = 50$ points and originates from the work of Petras and Podlubny [30]. It has been utilized also by Snow and Schaffrin, see Table A1 in [6], for a solution within the GH model, which will be used here for validating the results of the presented WTLS solutions. Three different stochastic models will be imposed in the following:

1. equal weights, i.e., coordinate components x_i, y_i, z_i as equally weighted and uncorrelated observations,
2. pointwise weights, i.e., coordinate components with same precision for each point and without correlations,
3. general weights, i.e., correlated coordinate components of individual precision including singular dispersion matrices.

3.1. Equal Weights

For the first case under investigation, we consider all coordinate components x_i, y_i, z_i as equally weighted and uncorrelated observations, yielding the weights as shown in (3). The least squares solution of this problem within the GH model, presented by Snow and Schaffrin [6], is listed in Table 1.

Table 1. Least squares solution within the Gauss–Helmert (GH) model of Snow and Schaffrin [6].

Line Orientation Components		Standard Deviation
Parameter $\hat{b}_y = a_n$	0.219309	0.077523
Parameter $\hat{b}_x = b_n$	0.677404	0.058450
Parameter $\hat{b}_z = c_n$	0.702159	0.056575
Coordinates of a point on the line		Standard deviation
Parameter $\hat{a}_y = y_0$	0.047785	0.121017
Parameter $\hat{a}_x = x_0$	−0.067111	0.091456
Parameter $\hat{a}_z = z_0$	0.049820	0.088503
A posteriori variance factor $\hat{\sigma}_0^2$	0.7642885	

A direct TLS solution for this problem can be derived using the approach presented in Section 2.1. The results are shown in Table 2. Numerically equal results have been derived by using the direct WTLS approach of Section 2.2 by setting all weights equal to one.

Table 2. Direct TLS solution from Section 2.1.

Line Orientation Components		Standard Deviation
Parameter a_n	0.219308632730	0.07752314583
Parameter b_n	0.677404488809	0.05844978733
Parameter c_n	0.702158730025	0.05657536189
A posteriori variance factor $\hat{\sigma}_0^2$	0.76428828602	

Comparing the solution with the one presented in Table 1, it can be concluded that the numerical results for the parameters coincide within the specified decimal places. Regarding the small difference in the numerical value for the variance factor, it is to be noted that the value in Table 2 was confirmed by two independent computations.

Furthermore, a point on the line can be easily computed using the equations of the functional model (2), as long as the direction vector parallel to the requested line is known. Alternatively, all the adjusted points will lie on the requested straight line, which can be simply computed by adding the computed residuals to the measured coordinates.

3.2. Pointwise Weights

For the second weighted case under investigation, we consider the coordinate components x_i, y_i, z_i of each point P_i to be uncorrelated and of equal precision. From the standard deviations listed in Table 3, the corresponding pointwise weights can be obtained from (11).

Table 3. Pointwise precision $\sigma_{xi} = \sigma_{yi} = \sigma_{zi} = \sigma_i$ for each point P_i.

Point i	σ_i	Point i	σ_i
1	0.802	26	0.792
2	0.795	27	0.799
3	0.807	28	0.801
4	0.770	29	0.807
5	0.808	30	0.798
6	0.799	31	0.796
7	0.794	32	0.792
8	0.808	33	0.806
9	0.807	34	0.805
10	0.800	35	0.801
11	0.789	36	0.808
12	0.798	37	0.778
13	0.808	38	0.795
14	0.803	39	0.794
15	0.804	40	0.803
16	0.808	41	0.772
17	0.806	42	0.791
18	0.806	43	0.806
19	0.807	44	0.804
20	0.806	45	0.807
21	0.804	46	0.803
22	0.808	47	0.808
23	0.805	48	0.801
24	0.801	49	0.805
25	0.801	50	0.779

A direct WTLS solution is derived, following the approach presented in Section 2.2. The determinant (23)

$$\begin{vmatrix} (d_1 - k) & d_4 & d_5 \\ d_4 & (d_2 - k) & d_6 \\ d_5 & d_6 & (d_3 - k) \end{vmatrix} = 0,$$

can be built with the components

$$\begin{aligned}
d_1 &= 213.505250528675, \\
d_2 &= 206.458905097029, \\
d_3 &= 198.273122927545, \\
\text{and} & \\
d_4 &= -20.9837621443375, \\
d_5 &= -12.1835697465792, \\
d_6 &= -81.6787394185243.
\end{aligned} \qquad (76)$$

which leads to the solutions for the Lagrange multiplier

$$\hat{k} = \begin{cases} 115.0596477492 & (\hat{k}_{\min}) \\ 218.3490470615 \\ 284.8285837425 \end{cases} \qquad (77)$$

The direct WTLS solution for the line orientation components is shown in Table 4.

Table 4. Direct WTLS solution from Section 2.2.

Line Orientation Components		Standard Deviation
Parameter a_n	0.230818543507	0.07646636344
Parameter b_n	0.677278360907	0.05781967335
Parameter c_n	0.698582007942	0.05623243170
A posteriori variance factor $\hat{\sigma}_0^2$	1.19853799739	

The presented results are numerically equal to the iterative WTLS solution using the algorithmic approach of Section 2.3, as well as the solution within the GH model.

3.3. General Weights

For the last weighted case in this investigation, we impose the most general case, i.e., correlated coordinate components with individual precision resulting in a singular dispersion matrix. To obtain such a matrix for our numerical investigations, we firstly solved the adjustment problem within the GH model with an identity matrix as dispersion matrix. From the resulting 150×150 dispersion matrix of the residuals

$$\mathbf{Q}_{vv} = \begin{bmatrix} \mathbf{Q}_{v_x v_x} & \mathbf{Q}_{v_x v_y} & \mathbf{Q}_{v_x v_z} \\ \mathbf{Q}_{v_y v_x} & \mathbf{Q}_{v_y v_y} & \mathbf{Q}_{v_y v_z} \\ \mathbf{Q}_{v_z v_x} & \mathbf{Q}_{v_z v_y} & \mathbf{Q}_{v_z v_z} \end{bmatrix}, \tag{78}$$

computed as e.g., presented by Malissiovas ([20], p. 46), we take the variances and covariances between the individual point coordinates, but not among the points, i.e., the diagonal elements of each sub-matrix in (78) and arrange them in a new 150×150 matrix

$$\mathbf{Q}_{LL} = \begin{bmatrix} \operatorname{diag}(\mathbf{Q}_{v_x v_x}) & \operatorname{diag}(\mathbf{Q}_{v_x v_y}) & \operatorname{diag}(\mathbf{Q}_{v_x v_z}) \\ \operatorname{diag}(\mathbf{Q}_{v_y v_x}) & \operatorname{diag}(\mathbf{Q}_{v_y v_y}) & \operatorname{diag}(\mathbf{Q}_{v_y v_z}) \\ \operatorname{diag}(\mathbf{Q}_{v_z v_x}) & \operatorname{diag}(\mathbf{Q}_{v_z v_y}) & \operatorname{diag}(\mathbf{Q}_{v_z v_z}) \end{bmatrix}, \tag{79}$$

with "diag()" denoting that only the diagonal elements are taken into account. This is an example of pointwise variances and covariances, described as case (i) in Section 2.3, but now yielding a singular dispersion matrix for the observations with

$$\operatorname{rank}(\mathbf{Q}_{LL}) = 100.$$

Before deriving an iterative WTLS solution for this weighted case, we must check if the criterion (66) for a unique solution of the adjustment problem is fulfilled. Therefore, we computed the 100×100 matrix $\mathbf{W}$ with

$$\operatorname{rank}(\mathbf{W}) = 100,$$

and the 100×4 matrix $\mathbf{A}$ with

$$\operatorname{rank}(\mathbf{A}) = 4.$$

The criterion ensures that a unique solution exists when using the presented singular dispersion matrix, while

$$\operatorname{rank}([\mathbf{W} \mid \mathbf{A}]) = 100, \tag{80}$$

since $n = 50$ observed points are used in this example, cf. (66). As for all iterative procedures, appropriate starting values for the unknowns must be provided. However, they can be obtained easily by first generating a direct solution with a simplified stochastic model. The iterative WTLS solution for the direction vector of the requested straight line is presented in Table 5.

Table 5. Iterative WTLS solution from Section 2.3.

Line Orientation Components		Standard Deviation
Parameter a_n	0.225471114499	0.076563026291
Parameter b_n	0.677670055415	0.057791104518
Parameter c_n	0.699947192665	0.056127005073
A posteriori variance factor $\hat{\sigma}_0^2$	0.798915322513	

The presented WTLS solution has been found to be numerically equal to the least squares solution within the GH model. Detailed numerical investigations of the convergence behaviour, e.g., in comparison to an adjustment within the GH model, are beyond the scope of this article. However, in many numerical examples it could be observed that the iterative WTLS approach showed a faster convergence rate compared to an adjustment within an iteratively linearized GH model.

4. Conclusions

For the problem of straight line fitting to 3D point data, two novel WTLS algorithms for two individual weighting schemes have been presented in this study:

- Direct WTLS solution for the case of pointwise weights, i.e., coordinate components with same precision for each point and without correlations,
- Iterative WTLS solution for the case of general weights, i.e., correlated coordinate components of individual precision including singular dispersion matrices. This algorithm works without linearizing the problem by Taylor series at any step of the solution process.

Both approaches are based on the work of Malissiovas [20], where similar algorithms have been presented for adjustment problems that belong to the same class, i.e., nonlinear adjustments that can be expressed within the EIV model. The approach presented in Section 2.1 provides a direct TLS solution assuming equally weighted and uncorrelated coordinate components. The fact that this assumption is inappropriate, e.g. for the evaluation of laser scanning data, has often been accepted in the past to provide a direct solution for large data sets. With the newly developed approach in Section 2.2 it is now possible to compute a direct WTLS solution at least for a more realistic stochastic model, namely pointwise weighting schemes.

If more general weight matrices must be taken into account in the stochastic model, including correlations or singular dispersion matrices, the presented algorithm of Section 2.3 can be utilized for an iterative solution without linearizing the problem by Taylor series at any step, following the algorithmic idea of WTLS. A criterion that ensures a unique solution of the problem when employing singular dispersion matrices has also been presented, which is based on the original ideas of Neitzel and Schaffrin [26,27], for a solution within the GH model.

Numerical examples have been presented in Section 3 for testing the presented WTLS algorithms. The utilized dataset of the observed 3D point data has also been employed by Snow and Schaffrin [6] for a solution of the problem within the GH model and originates from the study of Petras and Podlubny [30]. The results of the presented algorithms have been compared in all cases with existing solutions or the solutions coming from existing algorithms and have been found to be numerically equal.

Author Contributions: Conceptualization, G.M., F.N. and S.P.; methodology, G.M. and S.P.; software, G.M. and S.W.; validation, F.N. and S.P.; formal analysis, G.M.; investigation, G.M. and S.W.; data curation, G.M. and S.W.; writing—original draft preparation, G.M. and F.N.; writing—review and editing, G.M., F.N., S.P. and S.W.; supervision, F.N. All authors have read and agreed to the published version of the manuscript.

Funding: This research received no external funding.

Acknowledgments: The authors acknowledge support by the German Research Foundation and the Open Access Publication Fund of TU Berlin.

Conflicts of Interest: The authors declare no conflict of interest.

References

1. Jovičić, D.; Lapaine, M.; Petrović, S. Prilagođavanje pravca skupu točaka prostora [Fitting a straight line to a set of points in space]. *Geod. List.* **1982**, *36*, 260–266. (In Croatian)
2. Kahn, P.C. Simple methods for computing the least squares line in three dimensions. *Comput. Chem.* **1989**, *13*, 191–195. [CrossRef]
3. Drixler, E. *Analyse der Form und Lage von Objekten im Raum [Analysis of the Form and Position of Objects in Space]*; Deutsche Geodätische Kommission bei der Bayerischen Akademie der Wissenschaften (DGK): München, Germany, 1993; Volume C, No. 409. (In German)
4. Späth, H. Zur numerischen Berechnung der Trägheitsgeraden und der Trägheitsebene [Numerical calculation of the straight line of inertia and the plane of inertia]. *avn Allg. Vermess. Nachr.* **2004**, *111*, 273–275. (In German)
5. Kupferer, S. Verschiedene Ansätze zur Schätzung einer ausgleichenden Raumgeraden [Different approaches for estimating an adjusted spatial line]. *avn Allg. Vermess.-Nachr.* **2004**, *111*, 162–170. (In German)
6. Snow, K.; Schaffrin, B. Line Fitting in Euclidian 3D-Space. *Stud. Geophys. Geod.* **2016**, *60*, 210–227. [CrossRef]
7. Ghilani, C. *Adjustment Computations, Spatial Data Analysis*, 6th ed.; John Wiley and Sons, Inc.: Hoboken, NJ, USA, 2018.
8. Niemeier, W. *Ausgleichungsrechnung [Adjustment Computations]*, 2nd ed.; Walter de Gruyter: New York, NY, USA, 2008. (In German)
9. Perović, G. *Least Squares (Monograph)*; Faculty of Civil Engineering, University of Belgrade: Belgrade, Serbia, 2005.
10. Pope, A. Some pitfalls to be avoided in the iterative adjustment of nonlinear problems. In Proceedings of the 38th Annual Meeting of the American Society of Photogrammetry, Washington, DC, USA, 12–17 March 1972; pp. 449–477.
11. Lenzmann, L.; Lenzmann, E. Strenge Auswertung des nichtlinearen Gauss-Helmert-Modells [Rigorous solution of the nonlinear Gauss-Helmert-Model]. *avn Allg. Vermess. Nachr.* **2004**, *111*, 68–73. (In German)
12. Golub, G.; Van Loan, C. An analysis of the total least squares problem. *SIAM* **1980**, *17*, 883–893. [CrossRef]
13. Van Huffel, S.; Vandewalle, J. Algebraic Connections Between the Least Squares and Total Least Squares Problems. *Numer. Math.* **1989**, *55*, 431–449. [CrossRef]
14. Van Huffel, S.; Vandewalle, J. *The Total Least Squares Problem: Computational Aspects and Analysis*; SIAM: Philadelphia, PA, USA, 1991.
15. Malissiovas, G.; Neitzel, F.; Petrovic, S. Götterdämmerung over total least squares. *J. Geod. Sci.* **2016**, *6*, 43–60. [CrossRef]
16. Schaffrin, B.; Wieser, A. On weighted total least-squares adjustment for linear regression. *J. Geod.* **2008**, *82*, 415–421. [CrossRef]
17. Shen, Y.; Li, B.; Chen, Y. An iterative solution of weighted total least-squares adjustment. *J. Geod.* **2011**, *85*, 229–238. [CrossRef]
18. Amiri-Simkooei, A.; Jazaeri, S. Weighted total least squares formulated by standard least squares theory. *J. Geod. Sci.* **2012**, *2*, 113–124. [CrossRef]
19. Snow, K. Topics in Total Least-Squares within the Errors-In-Variables Model: Singular Cofactor Matrices and Prior Information. Ph.D. Thesis, Division of Geodetic Science, School of Earth Sciences, The Ohio State University, Columbus, OH, USA, 2012; Report. No. 502.
20. Malissiovas, G. *New Nonlinear Adjustment Approaches for Applications in Geodesy and Related Fields*; Ausschuss Geodäsie der Bayerischen Akademie der Wissenschaften (DGK): München, Germany, 2019; Volume C, No. 841.
21. Guo, C.; Peng, J.; Li, C. Total least squares algorithms for fitting 3D straight lines. *Int. J. Appl. Math. Mach. Learn.* **2017**, *6*, 35–44. [CrossRef]
22. Neitzel, F.; Petrovic, S. Total Least Squares (TLS) im Kontext der Ausgleichung nach kleinsten Quadraten am Beispiel der ausgleichenden Geraden [Total Least Squares in the context of least squares adjustment of a straight line]. *zfv Z. für Geodäsie, Geoinf. und Landmanagement* **2008**, *133*, 141–148. (In German)
23. Neitzel, F. Generalisation of Total Least-Squares on example of unweighted and weighted similarity transformation. *J. Geod.* **2010**, *84*, 751–762. [CrossRef]

24. Bronshtein, I.; Semendyayev, K.; Musiol, G.; Muehlig, H. *Handbook of Mathematics*, 5th ed.; Springer: Berlin/Heidelberg, Germnay; New York, NY, USA, 2007.
25. Borovička, J.; Spurný, P.; Keclíková, J. A new positional astrometric method for all-sky cameras. *Astron. Astrophys. Suppl. Ser.* **1995**, *112*, 173–178.
26. Neitzel, F.; Schaffrin, B. On the Gauss-Helmert model with a singular dispersion matrix where BQ is of smaller rank than B. *J. Comput. Appl. Math.* **2016**, *291*, 458–467. [CrossRef]
27. Neitzel, F.; Schaffrin, B. Adjusting a 2D Helmert transformation within a Gauss-Helmert model with a singular dispersion matrix where BQ is of smaller rank than B. *Acta Geod. Geophys. Montan. Hung.* **2017**, *52*, 479–496. [CrossRef]
28. Wells, D.; Krakiwsky, E.J. *The Method of Least Squares*; Lecture Notes 18; Department of Geodesy and Geomatics Engineering, University of New Brunswick: Fredericton, NB, Canada, 1971.
29. Mikhail, E.M. *Observations and Least Squares*; Harper & Row Publishers: New York, NY, USA; Hagerstown, MD, USA; San Francisco, CA, USA; London, UK, 1976.
30. Petras, I.; Podlubny, I. State space description of national economies: The V4 countries. *Comput. Stat. Data Anal.* **2007**, *52*, 1223–1233. [CrossRef]

© 2020 by the authors. Licensee MDPI, Basel, Switzerland. This article is an open access article distributed under the terms and conditions of the Creative Commons Attribution (CC BY) license (http://creativecommons.org/licenses/by/4.0/).

Article

Stochastic Properties of Confidence Ellipsoids after Least Squares Adjustment, Derived from GUM Analysis and Monte Carlo Simulations

Wolfgang Niemeier [1,*] and Dieter Tengen [2]

[1] Institut of Geodesy and Photogrammetry, Technische Universität Braunschweig, Bienroder Weg 81, 38106 Braunschweig, Germany
[2] Geotec Geodätische Technologien GmbH, 30880 Laatzen, Germany; dieter.tengen@geotec-gmbh.de
* Correspondence: w.niemeier@tu-bs.de; Tel.: +49-531-94573

Received: 30 June 2020; Accepted: 3 August 2020; Published: 8 August 2020

Abstract: In this paper stochastic properties are discussed for the final results of the application of an innovative approach for uncertainty assessment for network computations, which can be characterized as two-step approach: As the first step, raw measuring data and all possible influencing factors were analyzed, applying uncertainty modeling in accordance with GUM (Guide to the Expression of Uncertainty in Measurement). As the second step, Monte Carlo (MC) simulations were set up for the complete processing chain, i.e., for simulating all input data and performing adjustment computations. The input datasets were generated by pseudo random numbers and pre-set probability distribution functions were considered for all these variables. The main extensions here are related to an analysis of the stochastic properties of the final results, which are point clouds for station coordinates. According to Cramer's central limit theorem and Hagen's elementary error theory, there are some justifications for why these coordinate variations follow a normal distribution. The applied statistical tests on the normal distribution confirmed this assumption. This result allows us to derive confidence ellipsoids out of these point clouds and to continue with our quality assessment and more detailed analysis of the results, similar to the procedures well-known in classical network theory. This approach and the check on normal distribution is applied to the local tie network of Metsähovi, Finland, where terrestrial geodetic observations are combined with Global Navigation Satellite System (GNSS) data.

Keywords: GUM analysis; geodetic network adjustment; stochastic properties; random number generator; Monte Carlo simulation

1. Introduction

For decades, the quality concepts in geodesy have been based on classical statistical theory and generally accepted assumptions, such as the normal distribution of observations, possible correlation between observations and law of variance propagation. For the here discussed least squares adjustment, the variance–covariance matrix for the unknowns is considered to be the best representation for quality of results.

These considerations are the basis for standard quality measures for precision, such as standard deviation, mean square error, error or confidence ellipses and prerequisites for the derivation of reliability measures, as well as for more detailed methods such as congruency analysis.

With the advent of GUM, i.e., the "Guide to the Expression of Uncertainty in Measurement", see [1–3], which has found wide acceptance within the community of measuring experts in natural sciences, physics and mechanical engineering, we may ask whether or not the traditional concepts for quality assessment for geodetic adjustment results are still valid or rather should be replaced by GUM-related new measures.

In this paper, we will participate in this discussion and will study the statistical properties of adjustment results, presenting a new approach in which the variations of the network adjustment results are derived by Monte Carlo simulations, where the quality variability of the input observations is computed in a rigorous procedure based on the rules of GUM.

2. Quality Assessment in Classical Geodetic Adjustment

2.1. Functional Model

Within the established methodology (see e.g., [4,5]), quality assessment in geodetic network adjustment is based on the analysis of the covariance matrix Σ_{xx} of the final adjusted coordinates x. In most cases, the starting point for the adjustment process is the Gauss–Markov (GM) model, given by the functional model.

$$l_{(n,1)} + v_{(n,1)} = A_{(n,u)} \cdot \hat{x}_{(u,1)}, \tag{1}$$

which gives the functional relations between the observations l_i and the unknowns x_j in a linear/often linearized form. In Equation (1), l is the $(n, 1)$—vector of observations l_i, which are in most cases reduced observations after linearization. A is the (n, u)—coefficient or design matrix, known as the Jacobian matrix. The vector $x(u, 1)$ contains the parameters x_i in the adjustment problem, where here—without lack of generality—just coordinate unknowns are considered. The $(n, 1)$—vector of residuals v accounts for the possible inconsistencies between observations and unknowns.

2.2. Stochastic Model

The stochastic relations for and between the observations l_i are given by the (n, n)—covariance matrix Σ_{ll} of exactly those n. quantities l_i, that are used as input variables in the adjustment model, but see critical remarks in Section 2.4. According to mathematical statistics, the covariance matrix for these input variables is given by:

$$\Sigma_{ll} = \begin{bmatrix} \sigma_1^2 & \rho_{12}\sigma_1\sigma_2 & \cdots & \rho_{1n}\sigma_1\sigma_n \\ \rho_{21}\sigma_2\sigma_1 & \sigma_2^2 & \cdots & \rho_{2n}\sigma_2\sigma_n \\ \vdots & \vdots & \ddots & \vdots \\ \rho_{n1}\sigma_n\sigma_1 & \rho_{n2}\sigma_n\sigma_2 & & \sigma_n^2 \end{bmatrix}, \tag{2}$$

where the terms σ_i^2 represent the variance estimates for the input variable l_i, and the terms $\rho_{ij}\sigma_i\sigma_j$ are the covariances between variables l_i and l_j. The correlation coefficient ρ_{ij} between the input variables l_i and l_j is rarely known and therefore in most applications the stochastic model is reduced to a diagonal matrix, where correlations are no longer considered.

$$\Sigma_{ll(n,n)} = \begin{bmatrix} \sigma_1^2 & & & & \\ & \sigma_2^2 & & 0 & \\ & & \sigma_3^2 & & \\ & 0 & & \ddots & \\ & & & & \sigma_n^2 \end{bmatrix}. \tag{3}$$

Some literature exists to estimate correlation coefficients, where serial correlation, external influencing factors or neglected effects are considered to obtain adequate ρ_{ij} values. For GNSS observations [5], the application of correlation coefficients is standard practice, at least for 3D network blockwise correlations for (3,3) where coordinates or coordinate differences are considered.

A further step to simplify the stochastic model and the computational effort is the usage of identical values for a priori variances σ_i^2 for each type of observation (e.g., for directions, distances, height differences, coordinate differences). Often these simplifications are justified by the assumed minor influence of correlations on the coordinate estimates themselves.

2.3. Traditional Quality Assessment

For this common GM approach, the target function for a least squares adjustment is given by the well-known condition:

$$\Omega_\Sigma = v^T Q_{ll}^{-1} v = v^T P v, \tag{4}$$

where the variance–covariance matrix is split up:

$$\Sigma_{ll} = \sigma_0^2 Q_{ll}. \tag{5}$$

Here Q_{ll} is called the cofactor matrix of observations and σ_0^2 is the variance of unit weight, which can be used to carry out an overall test of the adjustment model, see e.g., [4]. For the final results of least squares adjustment, the coordinates (more precise: corrections to the approximate coordinates) are computed by the well-known formula:

$$\hat{x} = \left(A^T Q_{ll}^{-1} A\right)^{-1} A^T Q_{ll}^{-1} l. \tag{6}$$

The only stochastic variable in this equation is the vector of observations l; according to the law of variance propagation, the cofactor matrix Q_{xx} or the covariance matrix Σ_{xx} of the estimated parameters x can be derived easily:

$$Q_{xx} = \left(A^T Q_{ll}^{-1} A\right)^{-1}, \tag{7}$$

$$\Sigma_{xx} = \sigma_0^2 Q_{xx}. \tag{8}$$

This matrix Σ_{xx} contains all the information to estimate quality measures for the coordinates of a network, more precisely estimates for precision of the adjustment parameters. In most cases, the precision of point coordinates is computed and visualized by confidence ellipses for a certain confidence level. For the example of a local tie network in Finland, the 95% confidence ellipses for the final 3D coordinates are depicted in Figure 8 and discussed in Section 5.3.

To estimate quantities for reliability of observations and of coordinates, the cofactor matrix Q_{vv} for the residuals v has to be computed, which can be done in a straightforward way by applying the law of variance progagation to Equation (1). Even if aspects of reliability will not be discussed in this paper, it should be pointed out, additionally, that reliability measures are dependent on adequate covariance matrices Σ_{xx}.

2.4. Critisicm of Traditional Approach

Due to the modern electronic sensors, for the users it is almost impossible to obtain knowledge on relevant internal measuring processes and the already applied computational steps within the sensors. Therefore it is not sufficient to follow the classical concept to derive dispersion measures out of repeated observations only. As is the case nowadays, making a measurement is often identical with pushing a button, therefore it is obvious that these "observations" do not contain sufficient information on the real variability of the complete measuring data and processing chain. Besides, in general the variable environmental conditions and the ability of the measuring team are not taken into account. For the stochastic properties of observations, one can state that for the standard approach to develop a stochastic model, the following problems ought to be addressed:

(i) The set-up of appropriate values for variances σ_i^2 and correlations ρ_{ij} are based on:

- Experiences (of persons who carry out the observations and computations);
- Values given by the manufacturer of the instruments;

-   Results of repeated observations during the measuring campaign (e.g., out of three repeated sets of measurements for direction observations).

(ii) This selection does not consider in detail that the input variables for an adjustment are based on a multi-step preprocessing, e.g., for classical total station observations these steps consist of:

    -   Corrections due to atmospheric conditions (temperature, air pressure, air humidity);
    -   Centering to physical mark (horizontal centering, considering the instrument's height);
    -   Geometric reduction to coordinate system used;
    -   Application of calibration corrections.

(iii) The traditional stochastic model does not consider the real environmental conditions during the measuring campaign (rain, heavy wind, frost, etc.) and a possible influence due to the quality of the personal conditions (training status, physical wealth, stress, etc.).

It is almost impossible to consider all these influences in the a priori estimates for the variances and covariances in a rigorous way; therefore, it is left to the responsible person for data processing, in which way—if any—he/she includes these factors in the variance–covariance matrix (Equation (2)). With the application of the concepts of GUM—see following sections—one can overcome these shortages.

3. Uncertainty Modeling According to GUM

3.1. General Idea of GUM

In contrast to classical error analysis and quality assessment, the concept of this new GUM (Guide to the Expression of Uncertainty in Measurement) can be considered as a radical paradigm change. Within its realization, several new subtasks have to be solved as pre-analysis steps to get a complete uncertainty analysis according to this new concept.

As outlined in the last section, the traditional statistical concept, which derives dispersion measures and correlations out of repeated independent (!) observations, does not cover the complexity of today's measuring processes, see Figure 1.

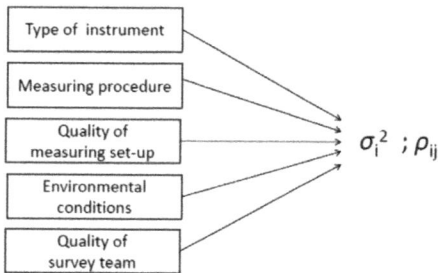

Figure 1. Set-up of a stochastic model within classical approach: external, instrumental and personal influences are considered "implicitly", at least in a subjective way.

Considering the deficiencies within the classical error theory, on initiative of the Bureau International des Poids et Mesures, France, an international group of experts of metrology formed in 1987 to develop a new approach to adequately assess the complete uncertainty budget of different types of measurements. As a result, the "Guide to the Expression of Uncertainty in Measurement" (GUM) was published, which nowadays is the international standard in metrology, see the fundamental publications [1,2]. The GUM allows the computation of uncertainty quantities for all measuring sensors or systems. The resulting uncertainty value is a non-negative parameter characterizing the complete dispersion of a measuring quantity and by this, nowadays, uncertainty values are considered to be the adequate precision parameters.

As described in the fundamental GUM documents, it is necessary to model the complete measuring and processing chain to derive a "final" resulting measuring quantity Y from all influencing raw data. It is important that this numerical model includes all, and really all, input quantities X_1, X_2, X_3 ..., that influence the final measuring result Y. As this model contains all the computational steps, including how the resulting quantity Y will be changed whenever one input quantity is modified, this basic model for GUM analysis is named carrier of information. In a simplified form, this model can be described as a (often nonlinear) complex function

$$Y = f(X_1, X_2, X_3, \ldots, X_n). \tag{9}$$

The development of this function is one of the most difficult and complex subtasks for deriving the uncertainty of measurements. Deep understanding of the physical and computational processes within the sensor, the performance of the measuring task itself, the data processing and possible external and environmental influences are necessary. To do this, no standard concept is available, just some recommendations can be given, see e.g., [6,7].

With respect to the later discussions here, it should be mentioned that the original GUM is going to derive an uncertainty measure for just one measuring quantity Y.

To restrict the contents of this paper, possible variabilities of the measurand—the physical quantity of interest, which is measured—is not considered here; as for this task, detailed physical knowledge of the specific object would be required.

3.2. Type A and Type B Influence Factors

An uncertainty analysis according to GUM has a probabilistic basis, but also aims to include all available knowledge of the possible factors that may influence the measuring quantity. Consequently, it is most important to set-up the following two types of influence factors, which are characterized as Type A and Type B:

Type A: Dispersion values for measurements

- Derived from common statistical approaches, i.e., analyzing repeated observations;
- Values following Gaussian distribution.

Type B: Non-statistical effects

- What are relevant external influences?
- Are there remaining systematic effects?
- Are insufficient formulas used during processing?
- Define possible variability within specified interval $[a, b]$;
- Assign a probability distribution to each influence factor.

The GUM concept allows us to consider classical random effects (Type A) on measuring results, which correspond to established statistical approaches.

However, additionally, GUM allows us to include all relevant additional influence factors (Type B), e.g., external effects (e.g., due to environmental conditions and the observing team) and possibly remaining systematic errors (e.g., uncontrolled residuals from the measuring procedure, undetected instrumental effects). Even approximations used in computational formulas have to be considered, and as such, are considered here.

3.3. Assignment of Adequate Probability Distribution Functions to Variables

The GUM concept requires the assignation of statistical distribution functions for all these influencing quantities of Type A and Type B, i.e., a specific distribution, its expectation and dispersion. This aspect is depicted in Figure 2.

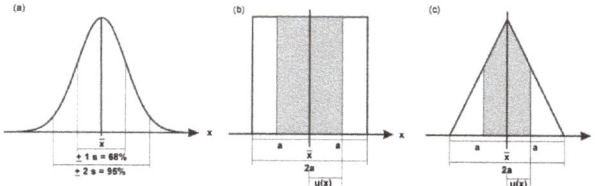

Figure 2. Statistical distribution functions, used within the Guide to the Expression of Uncertainty in Measurement (GUM) approach to model effects of Type A and Type B, from [8]. (**a**) Normal distribution, (**b**) uniform distribution and (**c**) rectangular distribution.

For Type A quantities, the common probability distribution functions with Gaussian or Normal distribution (with parameter expectation μ and variance σ^2) are applied, which is depicted in Figure 2a. Here, classical methods for variance estimation can be used, i.e., the statistical analysis of repeated measurements from our own or external experiences, adopt data sheet information, etc.

For the non-statistical influence factors of Type B, which represent external influences and remaining systematic effects as well as insufficient approximations, according to, e.g., [9,10], it is recommended to introduce a probability distribution function in addition. However, the individual assignment of an adequate statistical distribution is a particularly complex task; in general, the statistical concepts of normal, uniform and triangle distribution functions are used, see Figure 2 and examples in Table 1.

Table 1. Type A and Type B influencing factors, possible probability distribution functions and variability range for typical geodetic observations, taken from [11].

	Influence Factors	Distribution	Examples
Type A	**Total station** - horizontal directions - vertical distances - slope distances	 normal normal normal	 σ_h = 0.2 mgon σ_v = 0.3 mgon σ_d = 0.6 mm + 1 ppm
	Levelling - height differences	 normal	 $\sigma_{\Delta h}$ = 0.6 mm/$\sqrt{\text{km}}$
	GNSS - baselines $\Delta x, \Delta y, \Delta z$	 normal	 σ_Δ = 2 mm
Type B	**Pillar und centering** - centering direction - centering offset - Target center definition	 uniform triangle uniform	 [0, 360°] [0, 0.1 mm] σ_t = 0.1 mm
	Instrument and target height	uniform	[0, 0.2 mm]
	Calibration parameters - additional constant - scale factor	 normal normal	 σ_A = 0.5 mm σ_S = 0.2 ppm
	Atmospheric parameters - temperature - air pressure - air humidity	 uniform uniform uniform	 [0, 1 K] [0, 1 mbar] [0, 5%]

GNSS, Global Navigation Satellite System.

For each influence factor, a statistical distribution has to be defined with an expected mean and dispersion, i.e., all these quantities have to be pre-selected to serve as starting values for a complete GUM analysis. To be more specific, it is the engineers' task to estimate the variability of the applied temperature correction during the measuring period, to estimate a quantity for the centering quality, to evaluate the correctness of calibration parameters, etc.

3.4. Approach to Perform a GUM Analysis for Geodetic Observations

For Electronic Distance Measurements (EDMs), a common geodetic measuring technique, the processing steps according to Equation (9) are depicted in Figure 3. Be aware that here, the complete mathematical formulas are not given, just the specific computational steps are outlined.

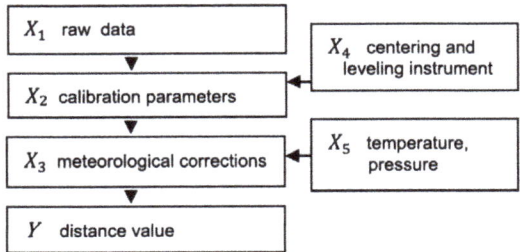

Figure 3. Influence factors for electronic distance measurements: processing steps for the derivation of an input quantity for an adjustment, from [11].

The resulting distance value Y, i.e., the numerical measuring quantity after all necessary pre-processing steps, will serve as the input quantity for network adjustment, see Section 4.2. Within the classical approach, it is necessary to assign a dispersion value to this quantity, see Section 2.3, but here, as an alternative, Monte Carlo simulations are applied.

A simplified numerical example for the set-up of Type A and Type B effects is given in Table 1, where the used geodetic measurements can be applied in a local 3D geodetic network. However, each project requires an individual evaluation of these more general reference values; note, for the numerical example in Section 5, we had to make slight changes of these reference values to account for specific measuring conditions.

The here listed influencing factors of Type A and Type B, as well as their corresponding probability distribution functions and domain of variability, do not claim to be complete, as they do not contain additional computational influences related to the reduction to a reference height (which is always required), effects to account for the selected surveying methods, the quality of the personal or the atmospheric and environmental influences, such as bad weather, strong insolation, etc.

The here presented selection of the distribution type and its variability range are solely preliminary steps. At minimum, a GUM analysis of GNSS observations, a much more detailed study of all influencing factors, has to be performed, which is a current project at the Finish Geodetic Institute [12].

The algorithmic complexity of the set-up of Equation (9), i.e., the difficulty to find the relevant carrier of information, makes it necessary to analyze the complete measuring process and all pre-processing steps. This problem can be visualized in a so-called Ishikawa diagram, as given in Figure 4. This frequently applied diagram, see [13], has to be filled out for each specific measurement system, which can be a laborious task, e.g., the actual publication [14].

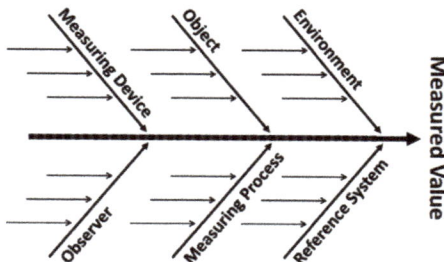

Figure 4. Ishikawa diagram to analyze all the influence factors for a GUM analysis.

3.5. Uncertainty Quantities out of GUM

To combine all these effects of Type A and Type B within the classical GUM approach, the well-known law of variance propagation is applied, despite the fact that these are effects with different probability distribution functions. In Figure 5, this approach is explained: On the left-hand side, the different Probability Distribution Functions (PDF) are visualized, i.e., normal, uniform and triangle distribution. On the right-hand side, the formula for the law of variance propagation is shown, combining different uncertainties u_{xi} of Type A and Type B. Of course, for each influence factor an individual element u_{xi} has to be considered.

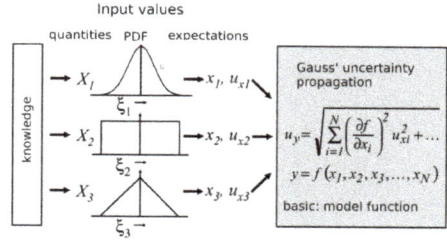

Figure 5. Classical concept to combine Type A and Type B influence parameters within GUM [7].

Numerically therefore, the uncertainty for the final measuring quantity Y, see Equation (9), is derived by combining the uncertainties of Type A (u_{Ai}) and Type B (u_{Bi}) for all influence factors following the formula:

$$u_Y = \sqrt{u_{A1}^2 + \ldots + u_{An}^2 + u_{B1}^2 + \ldots + u_{Bm}^2}. \tag{10}$$

Within GUM an extended and a complete uncertainty is introduced as well, both derived quantities out of u_Y. A discussion of the usefulness of these extensions and their computations is outside the scope of this paper.

3.6. Criticism

The application of statistical distribution functions to Type B errors and the application of the law of variance propagation to obtain an uncertainty estimate u_Y are critical points within the GUM approach [15]. The assignment of probability distribution functions to the influencing factors is a sensitive step and of course, the application of the law of variance propagation is a practical method, but it allows us to stay with the established statistical methods and perform subsequent computations.

This GUM concept is discussed within recent geodetic literature to some extent, see e.g., [13–17]. However, most of these discussions and critical remarks are limited to an uncertainty assessment for single measurements, not for complex networks or systems.

Taking into account this criticism, in a later extension of the GUM concept [2,3,18], the use of Monte Carlo simulations is recommended to find the final distribution for the quantity Y.

We will not follow this concept, but extend the processing model, see Figure 6, to directly study the stochastic properties of the outcome of the least squares adjustment. Due to our knowledge, such a network approach has not been considered yet.

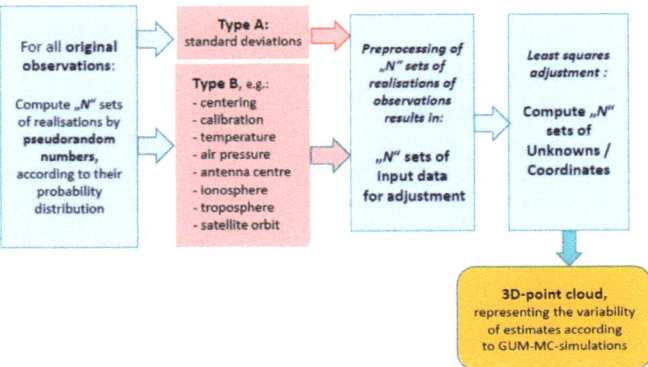

Figure 6. GUM concept for creation of "N" sets of input data used in Monte Carlo (MC) simulations for least squares network adjustments.

4. Monte Carlo Simulations

4.1. Basic Idea of MC Simulations

For decades, Monte Carlo (MC) methods have been developed and in use to solve complex numerical problems in a specific way, i.e., by repeated random experiments, performed on a computer, see e.g., [19]. All MC computations use repeated random sampling for input quantities, process these input data according to the existing algorithms and obtain a variability of numerical results. For typical simulations, the repeat rate of experiments is 1000–100,000 or more, in order to obtain the most probable distribution of the quantities of interest. This allows MC simulations to model phenomena with well-defined variability ranges for input variables.

Nowadays, with modern computers and well-established Random Number Generators (RNG), large samples are easy to generate, see [20,21]. According to the variability of input quantities, pseudorandom sequences are computed, which allows us to evaluate and re-run simulations.

In this paper, MC simulations are applied to perform uncertainty modelling according to GUM, the specific case of geodetic data processing, i.e., network adjustment following a traditional Gauss–Markov (GM) model. The use of MC simulations allows us to include different Type A and Type B influence factors, which is an extension in relation to the classical approach. The approach allows us to combine a detailed GUM analysis of the measurement process with MC simulations in a rigorous way.

The typical pattern of an MC simulation is as follows:

- Define functional relations between all input data and the quantities of interest.
- Define probability distribution functions and variability ranges for starting data.
- Generate corresponding input data with RNG.
- Perform deterministic computations with these input values and obtain a pre-set number of realizations for the quantities of interest.
- Analyze the achieved quantities of interest.

4.2. Concept to Combine MC-Simulations with GUM Analysis

The scheme for the here proposed approach for an uncertainty assessment within least squares adjustments of geodetic networks by a rigorous combination of MC simulations with GUM analysis is presented in Figure 6. Starting point is an analysis of the complete pre-processing chain for each observation l_i according to GUM, i.e., an analysis of all possible influencing factors of Type A and Type B, according to the important carrier of information formula, see Equation (9). For all these influencing factors, the most probable numerical measuring value is the starting point, often a mean value or a real observation.

As the next step, pseudorandom numbers for all influencing factors for each observation are created, taking into account their most probable value, the selected probability distribution function and the variability domain. There are numerous options for selecting a Random Number Generator (RNG). RNG should have high quality, i.e., provides good approximations of the ideal mathematical system, e.g., has long sequences, shows no gaps in data, fulfils distribution requirements, see [21]. As discussed in [20], the RANLUX (random number generator at highest luxury level) and its recent variant RANLUX++, which are used here, can be considered as representative of such high-quality RNGs.

For each original reading, respectively, for each influence factor, by using this RNG, a random value is created, representing one realization of the real measuring process. By combining these effects in a consecutive way, see the simplified example in Table 2, for each input quantity for network adjustment, such a randomly generated value is gained. With each set of input data, one least squares adjustment is performed, coming up with one set of coordinate estimates as the outcome.

Table 2. Derivation of one input quantity for a distance, using random numbers for some influencing effects.

Action	Stochastic Properties	Resulting in Random Distance Value:
"True" coordinates		$x_1 = 100.0000$ m, $y_1 = 100.0000$ m $x_2 = 200.0000$ m, $y_2 = 200.0000$ m
+ Pillar variations result in: "real distance"	Uniform distribution: $\sigma = 0.1$ mm	$x_i = 100.00005$ m, $y_i = 99.99999$ m $x_2 = 199.99998$ m, $y_2 = 200.00001$ m $S^r = 141.42132$ m
+ Calibration effects additional constant: scale	Normal distribution: $\sigma = 0.5$ mm Normal distribution: $\sigma = 1$ ppm	$S^{rc} = 141.42165$ m
+ Weather effects temperature air pressure	Uniform distribution: $\sigma = 1$ K Uniform distribution: $\sigma = 5$ mbar	$S^{rcw} = 141.42163$ m
+ Type A uncertainties Constant effect Distance dependent	Normal distribution: $\sigma = 0.6$ mm Normal distribution: $\sigma = 1$ ppm	$S^{rcwd} = 141.42136$ m

Repeating this complete approach for a preset number of N (e.g., 1000 or 10,000) simulations, the final results of a GUM-MC simulation are achieved, i.e., a set of coordinates/unknowns with its variability, which represent the uncertainty of the coordinates according to the used GUM analysis.

As an example, the specific manner, in which the random numbers for distance observations as input quantities for the adjustment are derived, are depicted in Table 2. Starting with a most probable mean value, such as Type B errors, the effects of pillar variations, calibration and weather and the classical Type A errors are considered. In column 2 their stochastic properties are given, which are the basis for the generation of a random number, which results in a specific modification of the distance observation, see column 3.

5. Application to a Local 3D Geodetic Network

5.1. Test Site "Metsähovi"

The Metsähovi Fundamental Station belongs to the key infrastructure of Finnish Geospatial Research Institute (FGI). Metsähovi is a basic station for the national reference system and the national permanent GNSS network. This station is a part of the global network of geodetic core stations, used to maintain global terrestrial and celestial reference frames as well as to compute satellite orbits and perform geophysical studies.

Of special interest here is the character of this network to serve as "Local Tie-Vector", see [22], which is defined as a 3D-coordinate difference between the instantaneous phase centers of various space-based geodetic techniques, in this case between VLBI (Very Long Baseline Interferometry), GNSS (Global Navigation Satellite System) and SLR (Satellite Laser Ranging). All these techniques have different instruments on the site and the geometric relations between their phase centers have to be defined with extreme precision.

As depicted in Figure 7, the structure of this network is rather complex, a special difficulty is that the VLBI antenna is located inside a radome. This requires a two-step network with a connection between an outside and inside network, which is the most critical part in the network design, but this will not be discussed here in detail.

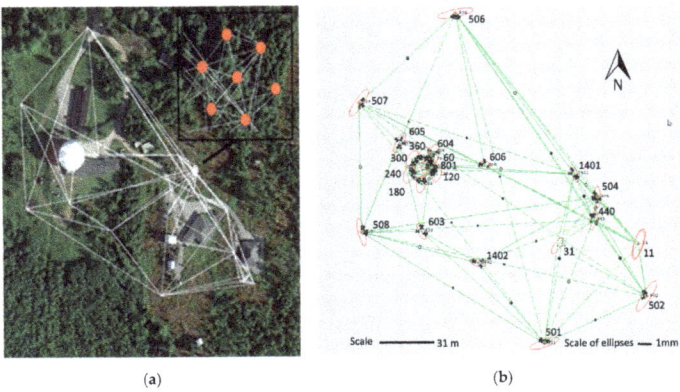

Figure 7. (a) Fundamental station Metsähovi, Finland, WGS84 (60.217301 N, 24.394529 E); with local tie network. (b) Network configuration with stations 11, 31 and 180, which define local tie vectors.

As already discussed in [11], the local tie network Metsähovi consists of 31 points, where specific local tie vectors are given by the GNSS stations, 11 and 31, on the one hand side and point 180, which is located within a radome. This station 180 is not the reference point of the VLBI antenna, it is located in its neighborhood. Here, a 3D free network adjustment is performed with the following measurement elements: 149 total station measurements (slope distances, horizontal directions and vertical angles), 48 GNSS baselines and levelled 43 height differences.

5.2. Input Variables for GUM Analysis

To get a realistic idea of uncertainties within this network, a classical adjustment model, which combines GNSS, total station and levelling measurements, is set up for this Metsähovi network. Then, all (many) influence factors are considered according to a GUM analysis. The Monte Carlo simulation process starts with the generation of pseudo random numbers for the influence factors, resulting in a set of input values for the adjustment. Repeating this MC simulation with 1000 runs gives the here discussed results.

The GNSS uncertainty model is just a rough idea, as in general it is difficult to simulate all influence factors with e.g., orbital errors, remaining atmospheric effects, multipath and near field effects. Colleagues from FGI are working on the problem to develop a more realistic GUM model for GNSS, see [12].

In our approach, the local tie network is simulated with 1000 runs of pre-analysis and least squares adjustment. In each run, a new set of observations is generated and subsequently a new set of station coordinates is computed. A forced centering is assumed, therefore, in each simulation the coordinates may differ only according to possible pillar centering variations.

The local tie vector consists of the outside stations 11 and 31 and station 180 in the radome, see above. The uncertainty estimates for these stations will be considered here in detail.

According to the GUM concept, the following influencing factors were considered. As mentioned in Section 3.4, some changes of the reference values in Table 1 were necessary to account for specific measurement conditions.

5.2.1. Type A: Classical Approach, Standard Deviations

Total station observations

As standard deviation for modern total stations (e.g., Leica TS30) often values of 0.15 mgon for manual angle measurements and of 0.3 mgon for observations with automatic target recognition are used. With two sets of angle observations, a precision of 0.2 mgon is assumed to be valid:

- Horizontal directions: normal distribution, $\sigma = 0.2$ mgon;
- Zenith angles: normal distribution, $\sigma = 0.2$ mgon;
- Slope distances: normal distribution, $\sigma = 0.6$ mm + 1 ppm.

GNSS Baselines

In these computations just a rough estimate is used, neglecting correlations between baseline components:

- For each coordinate component: normal distribution, $\sigma = 3.0$ mm.

Height differences

- Height differences: normal distribution, $\sigma = 2.0$ mm/$\sqrt{km}$.

5.2.2. Type B: Additional Influences, including Systematic Effects, External Conditions, Insufficient Approximations, etc.

Variation of pillar and centering

- Uniform distribution, range: 0–0.1 mm.

Variation of instrument and target height

- Uniform distribution, range: 0–0.1 mm.

Effects of calibration (total station instrument)

Schwarz [14] gives possible standard deviations of calibration parameters for total stations:

- Additive constant: normal distribution, $\sigma = 0.2$ mm (valid for combination of instrument and specific prism).
- Scale unknown: normal distribution, $\sigma = 0.8$ ppm. The value for scale is related to the problem to determine a representative temperature along the propagation path of the laser beam.

Effect of calibration of GNSS antenna

- Not implemented yet because these effects were not known to us for the test site Metsähovi.

Effects of limited knowledge on atmospheric parameters

- Air temperature: uniform distribution, range 0–0.8 K; (effect: 1 K ≈ 1 ppm);
- Air pressure: uniform distribution, range 0–0.5 mbar; (effect: 1 mbar ≈ 0.3 ppm);
- Air humidity: uniform distribution, range 0–5%.

5.3. Final Results of GUM-Analysis and MC Simulations

We applied to this network the developed approach of detailed GUM analysis and full MC simulation, i.e., starting with a simulation of the original influence factors and then performing an adjustment. In Figures 8 and 9, the resulting point clouds are depicted for coordinates for stations 11, 31 and 180, which form the local tie vectors. As the Metsähovi network is a 3D geodetic network, for simplicity the resulting point clouds are visualized in the X–Y plane and X–Z plane.

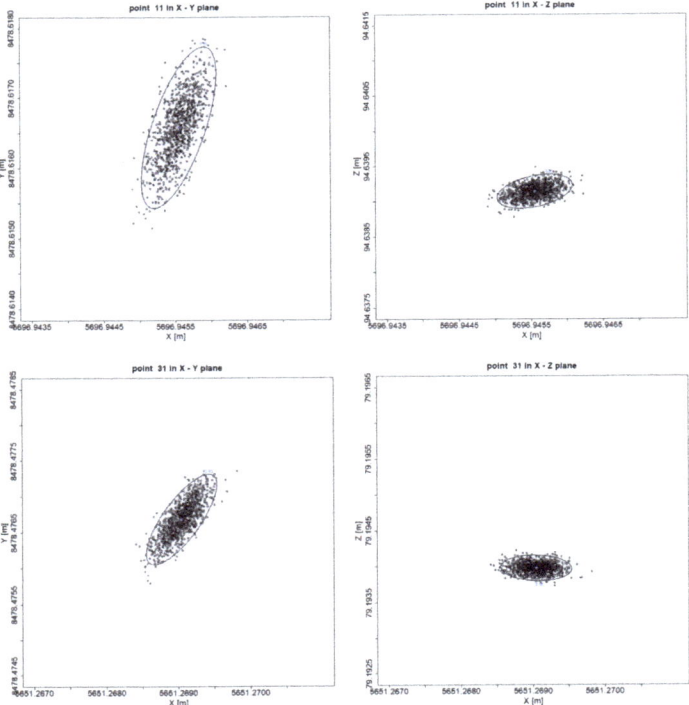

Figure 8. *Cont.*

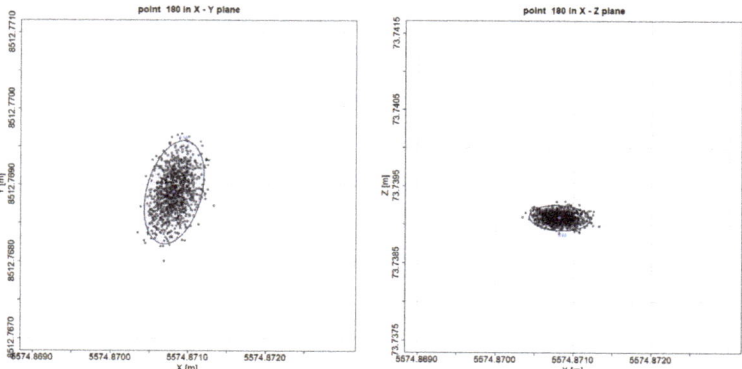

Figure 8. Point clouds of coordinate variations for stations 11, 31 and 180. **Left:** in X-Y plane. **Right:** in X-Z plane. The elliptical contour lines refer to a confidence level of 95%.

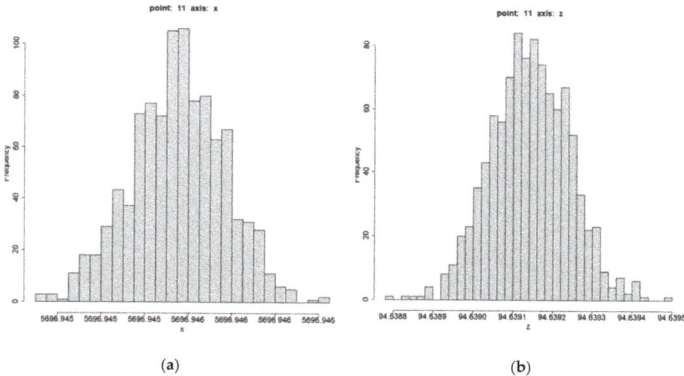

Figure 9. Histograms of the variations of (**a**) x- and (**b**) z-coordinates of station 11 after GUM and MC simulation.

This variation of all coordinate components is of special interest here, as the local tie vectors are defined between the GPS Reference stations 11 and 31 and station 180 inside the radome. The original coordinate differences refer to the global cartesian coordinate system.

6. On Stochastic Properties of These Point Clouds

The focus of this paper is the stochastic properties for the results of such a MC simulation where the input data are randomly variated quantities, following the GUM analysis concept, see the scheme in Figure 6 and the description given in Section 4.2.

For this analysis, we consider the point clouds for final coordinates as information of primary interest, as depicted e.g., in Figures 8 and 9, separated into X-Y and X-Z planes to make it easier to visualize the findings. To study the stochastic properties of these results, two concepts from mathematical statistics can be considered.

6.1. Cramér's Central Limit Theorem

Already in 1946, the statistician Cramér [23] developed a probabilistic theory, which is well-known as the Central Limit Theorem (CLT). Without going into detail, the general idea of the CLT is, that the properly normalized sum of independent random variables tends towards a normal distribution, when they are added. This theorem is valid, even if it cannot be guaranteed that the original variables are normally distributed. This CLT is a fundamental concept within probability theory because it allows for the application of probabilistic and statistical methods, which work for normal distributions, to many additional problems, which involve different types of probability distributions.

Following Section 4.2, during the GUM analysis a number of influence factors with different distributions are combined, what is of course more than the "properly normalized sums", which are asked for in the CLT. However, the question is whether or not the resulting input variables of the adjustment are sufficiently normal distributed or—considered here—if the resulting point clouds after MC simulations for adjustment using these input datasets tend to have a normal distribution.

6.2. Hagen's Elementary Error Theory

Within geodesy and astronomy, a theory of elementary errors was already developed by Hagen in 1837 [24]. This concept means that randomly distributed residuals ε, for which we assume the validity of a normal distribution, can be considered to be the sum of a number of q very small elementary errors Δ_i:

$$\varepsilon = \Delta_1 + \Delta_2 + \Delta_3 + \Delta_4 + \ldots + \Delta_q. \tag{11}$$

This equation holds—according to Hagen's discussion—if all elementary errors have similar absolute size and positive and negative values have the same probability.

Of course the conditions for this theory are not really fulfilled by the here considered concept of GUM analysis of influence factors with subsequent MC simulation, but out of this elementary error theory one could ask whether or not the results of the here proposed approach tends to have a normal distribution, as well.

6.3. Test of MC Simulation Results on Normal Distribution

To analyze the statistical distribution of a sample of data, numerous test statistics are developed. For analyzing the normal distribution, here we used the test statistics of Shapiro–Wilks, Anderson–Darling, Lilliefors (Kolmogorov–Smirnov) and Shapiro–Francia. A detailed description of these test statistics is given in the review article [25]. The principle of all these tests is to compare the distribution of empirical data with the theoretical probability distribution function.

Here, we want to perform one-dimensional tests for the variation of the x-, y- and z-components of the stations 11, 31 and 180 of our reference network Metsähovi. The used coordinates are the results of 1000 MC simulations, where the input data are pre-processed according to the concept of Section 4.2.

The above-mentioned numerical tests for these simulated values accept for all stations the hypothesis of normal distribution. For us, this is an indication that should be allowed in order to consider the results of the here proposed approach of having a normal distribution. This means an uncertainty for station coordinates, derived via GUM analysis and subsequent MC simulation, can be treated for ongoing statistical analysis of results, such as general quality assessment and deformation analysis, in the same way as the results of classical adjustment. Some additional graphical information for these comparisons can be given, as well. In Figure 9, the distribution of the x- and z-components of station 11 are depicted, received from least squares adjustments after GUM and MC simulations, i.e., corresponding the results of Figure 8. It is obvious that the appearance of the components is close to the well-known bell curve of normal distribution. For the stations 31 and 180 these distributions were visualized, as well, with very similar appearances.

The Q–Q (quantile–quantile) plot is a graphical method for comparing two probability distributions by representing their quantities in a plot against each other. If the two distributions being compared are similar, the points in the Q–Q plot will approximately appear in a line.

If the most real points are within the dashed line, here representing a normal distribution, the assumption of a normal distribution is allowed. The confidence level for point-wise confidence envelope is 0.95. From Figure 10, it can be seen that for station 180 a normal distribution can be accepted. For stations 11 and 31, the assumption of a normal distribution is accepted, as well.

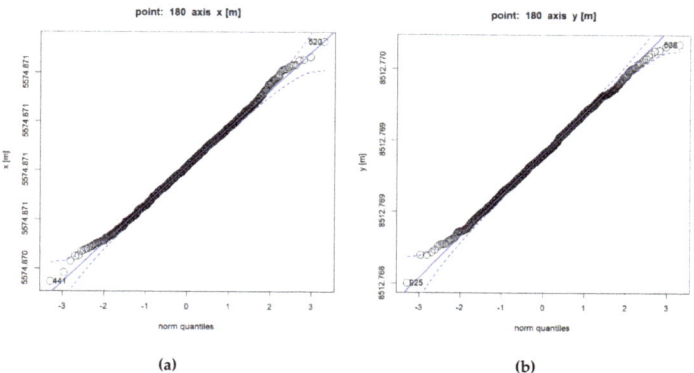

Figure 10. Q–Q plots for comparing probability distributions of (**a**) x- and (**b**) y-coordinates of station 180: If real values lie within dashed line, here assumed normal distribution is accepted.

7. Conclusions

Within the classical approach of network adjustment, see e.g. [4,5], more or less rough and subjective estimates for the dispersion of measurements are introduced. A detailed criticism to this approach is given in Section 2.4. The here described GUM-based approach starts with an analysis of the complete measuring process and computational process, i.e., it considers all influence factors, which are used during the pre-processing steps. By this, the GUM approach can be considered to give more realistic quantities for the precision of the input data of an adjustment of geodetic networks.

Using random numbers to cover the variability of these input data allows us to perform Monte Carlo simulations, which makes it possible to compute the corresponding variability or uncertainty ranges of final coordinate sets.

These variations are analyzed statistically and tend to follow a normal distribution. This allows us to derive confidence ellipsoids out of these point clouds and to continue with classical quality assessment and more detailed analysis of results, as is performed in established network theory.

Of course, these results are dependent on the selection of the statistical distributions and their variability domains during the GUM analysis of the relevant Type A and Type B influence factors. This concept allows a new way of analyzing the effects of all influence factors on the final result, i.e., the form and size of derived ellipsoids. For example, one can analyze the effect of a less precise total station or a better GNNS system the same way one can study the influence of limited knowledge of the atmospheric conditions or a more rigorous centering system. These studies are outside the scope of this paper. Anyway, the here proposed approach allows a straightforward application of the GUM concept to geodetic observations and to geodetic network adjustment. Being optimistic, the here presented concept to derive confidence ellipsoids out of GUM–MC simulations could replace the classical methods for quality assessment in geodetic networks. The GUM approach will lead to uncertainty estimates, which are more realistic for modern sensors and measuring systems and it is about time to adapt these new concepts from metrology within the discipline of geodesy.

Author Contributions: Conceptualization, methodology and writing: W.N.; formal analysis, software development, visualization and review: D.T. Both authors have read and agreed to the published version of the manuscript.

Funding: Preliminary work for this project is performed within the frame work of the joint research project SIB60 "Surveying" of the European Metrology Research Programme (EMRP). EMRP research is jointly funded by the participating countries within EURAMET and the European Union. The analysis of stochastic properties, presented here, has not received external funding.

Acknowledgments: The authors have to thank the company Geotec Geodätische Technologien GmbH, Laatzen, Germany, for allowing us to use a modification of their commercial software package PANDA for carrying out all computations presented here.

Conflicts of Interest: The authors declare no conflict of interest.

References

1. JCGM (100:2008): Evaluation of Measurement Data—An Introduction of the "Guide to the Expression of Uncertainty in Measurements". Available online: https://www.bipm.org (accessed on 31 March 2020).
2. JCGM (101:2008): Evaluation of Measurement Data-Supplement 1 "Guide to the Expression of Uncertainty in Measurement" Propagation of Distributions Using a Monte Carlo Method. Available online: https://www.bipm.org (accessed on 2 April 2020).
3. Bich, W. Uncertainty Evaluation by Means of a Monte Carlo Approach. In Proceedings of the BIPM Workshop 2 on CCRI Activity Uncertainties and Comparisons, Sèvres, France, 17–18 September 2008; Available online: http://www.bipm.org/wg/CCRI(II)/WORKSHOP(II)/Allowed/2/Bich.pdf (accessed on 21 April 2020).
4. Niemeier, W. *Ausgleichungsrechnung*, 2nd ed.; Walter de Gruyter: Berlin, Gemany, 2008.
5. Strang, G.; Borre, K. *Linear Algebra, Geodesy and GPS*; Wellesley-Cambridge Press: Wellesley, MA, USA, 1997.
6. Kessel, W. Measurement uncertainty according to ISO/BIPM-GUM. *Thermochim. Acta* **2002**, *382*, 1–16. [CrossRef]
7. Sommer, K.-D.; Siebert, B. Praxisgerechtes Bestimmen der Messunsicherheit nach GUM. *Tm Tech. Mess.* **2004**, *71*, 52–66. [CrossRef]
8. Meyer, V.R. Measurement Uncertainty. *J. Chromatogr. A* **2007**, *1158*, 15–24. [CrossRef] [PubMed]
9. Sivia, D.S. *Data Analysis-A Bayesian Tutorial*; Clarendon Press: Oxford, UK, 1996.
10. Weise, K.; Wöger, W. *Meßunsicherheit und Meßdatenauswertung*; Wiley-VCH Verlag GmbH & Co. KGaA: Weinheim, Germany, 1999.
11. Niemeier, W.; Tengen, D. Uncertainty assessment in geodetic network adjustment by combining GUM and Monte-Carlo-simulations. *J. Appl. Geod.* **2017**, *11*, 67–76. [CrossRef]
12. Kallio, U.; Koivula, H.; Lahtinen, S.; Nikkonen, V.; Poutanen, M. Validating and comparing GNSS antenna calibrations. *J. Geod.* **2019**, *93*, 1–18. [CrossRef]
13. Hennes, M. Konkurrierende Genauigkeitsmasse—Potential und Schwächen aus sicht des Anwenders. *Allg. Vermess. Nachr.* **2007**, *114*, 136–146.
14. Schwarz, W. Methoden der Bestimmung der Messunsicherheit nach GUM-Teil 1. *Allg. Vermess. Nachr.* **2020**, *127*, 69–86.
15. Kutterer, H.-J.; Schön, S. Alternativen bei der Modellierung der Unsicherheiten beim Messen. *Z. Vermess.* **2004**, *129*, 389–398.
16. Neumann, I.; Alkhatib, H.; Kutterer, H. *Comparison of Monte-Carlo and Fuzzy Techniques in Uncertainty Modelling*; FIG Symposium on Deformation Analysis: Lisboa, Portugal, 2008.
17. Jokela, J. *Length in Geodesy—On Metrological Traceability of a Geospatial Measurand*; Publications of the Finnish Geodetic Institute: Helsinki, Finland, 2014; Volume 154.
18. Siebert, B.; Sommer, K.-D. Weiterentwicklung des GUM und Monte-Carlo-Techniken. *Tm Tech. Mess.* **2004**, *71*, 67–80. [CrossRef]
19. Kroese, D.P.; Taimre, T.; Botev, Z.I. *Handbook of Monte Carlo Methods*; Wiley & Sons: Hoboken, NJ, USA, 2011; Volume 706.
20. James, F.; Moneta, L. Review of High-Quality Random Number Generators. In *Computing and Software for Big Science*; Springer Online Publications: Berlin/Heidelberg, Germany, 2020.
21. Knuth, D.E. *The Art of Computer Programming, Volume 2: Semi-Numerical Algorithms*, 3rd ed.; Addison-Wesley: Reading, PA, USA, 1998.
22. Ning, T.; Haas, R.; Elgered, G. Determination of the Telescope Invariant Point and the local tie vector at Onsala using GPS measurements. In *IVS 2014 General Meeting Proceedings "VGOS: The New VLBI Network"*; Science Press: Beijing, China, 2014; pp. 163–167.

23. Cramér, H. Mathematical models in statistics. In *Princeton Math. Series*, 9th ed.; Princeton University Press: Princeton, NJ, USA, 1946.
24. Hagen, G. *Grundzüge der Wahrscheinlichkeitsrechnung*; Verlag von Ernst&Korn: Berlin, Germany, 1837.
25. Ogunleye, L.I.; Oyejola, B.A.; Obisesan, K.O. Comparison of Some Common Tests for Normality. *Int. J. Probab. Stat.* **2018**, *7*, 130–137.

© 2020 by the authors. Licensee MDPI, Basel, Switzerland. This article is an open access article distributed under the terms and conditions of the Creative Commons Attribution (CC BY) license (http://creativecommons.org/licenses/by/4.0/).

Article

Mean Shift versus Variance Inflation Approach for Outlier Detection—A Comparative Study

Rüdiger Lehmann [1,*], Michael Lösler [2] and Frank Neitzel [3]

1. Faculty of Spatial Information, University of Applied Sciences Dresden, 01069 Dresden, Germany
2. Faculty 1: Architecture—Civil Engineering—Geomatics, Frankfurt University of Applied Sciences, 60318 Frankfurt, Germany; michael.loesler@fb1.fra-uas.de
3. Technische Universität Berlin, Institute of Geodesy and Geoinformation Science, 10623 Berlin, Germany; frank.neitzel@tu-berlin.de
* Correspondence: ruediger.lehmann@htw-dresden.de; Tel.: +49-351-462-3146

Received: 8 May 2020; Accepted: 12 June 2020; Published: 17 June 2020

Abstract: Outlier detection is one of the most important tasks in the analysis of measured quantities to ensure reliable results. In recent years, a variety of multi-sensor platforms has become available, which allow autonomous and continuous acquisition of large quantities of heterogeneous observations. Because the probability that such data sets contain outliers increases with the quantity of measured values, powerful methods are required to identify contaminated observations. In geodesy, the mean shift model (MS) is one of the most commonly used approaches for outlier detection. In addition to the MS model, there is an alternative approach with the model of variance inflation (VI). In this investigation the VI approach is derived in detail, truly maximizing the likelihood functions and examined for outlier detection of one or multiple outliers. In general, the variance inflation approach is non-linear, even if the null model is linear. Thus, an analytical solution does usually not exist, except in the case of repeated measurements. The test statistic is derived from the likelihood ratio (LR) of the models. The VI approach is compared with the MS model in terms of statistical power, identifiability of actual outliers, and numerical effort. The main purpose of this paper is to examine the performance of both approaches in order to derive recommendations for the practical application of outlier detection.

Keywords: mean shift model; variance inflation model; outlierdetection; likelihood ratio test; Monte Carlo integration; data snooping

1. Introduction

Nowadays, outlier detection in geodetic observations is part of the daily business of modern geodesists. As Rofatto et al. [1] state, we have well established and practicable methods for outlier detection for half a century, which are also implemented in current standard geodetic software. The most important toolbox for outlier detection is the so-called data snooping, which is based on the pioneering work of Baarda [2]. A complete distribution theory of data snooping, also known as DIA (detection, identification, and adaptation) method, was developed by Teunissen [3].

In geodesy, methods for outlier detection can be characterised as statistical model selection problem. A null model is opposed to one or more extended or alternative models. While the null model describes the expected stochastic properties of the data, the alternative models deviate from such a situation in one way or another. For outlier detection, the alternative models relate to the situation, where the data are contaminated by one or more outliers. According to Lehmann [4], an outlier is defined by "an observation that is so probably caused by a gross error that it is better not used or not used as it is".

From a statistical point of view, outliers can be interpreted as a small amount of data that have different stochastic properties than the rest of the data, usually a shift in the mean or an inflation of the variance of their statistical distribution. This situation is described by extra parameters in the functional or stochastic model, such as shifted means or inflated variances. Such an extended model is called an alternative model. Due to the additionally introduced parameters, the discrepancies between the observations and the related results of the model decrease w. r. t. the null model. It has to be decided whether such an improvement of the goodness of fit is statistically significant, which means that the alternative model describes the data better than the null model. This decision can be made by hypothesis testing, information criteria, or many other statistical decision approaches, as shown by Lehmann and Lösler [5,6].

The standard alternative model in geodesy is the mean shift (MS) model, in which the contamination of the observations by gross errors is modelled as a shift in the mean, i.e., by a systematic effect. This approach is described in a large number of articles and textbooks, for example, the contributions by Baarda [2], Teunissen [7], and Kargoll [8]. However, there are other options besides this standard procedure. The contamination may also be modelled as an inflation of the variance of the observations under consideration, i.e., by a random effect. This variance inflation (VI) model is rarely investigated in mathematical statistics or geodesy. Bhar and Gupta [9] propose a solution based on Cook's statistic [10]. Although this statistic was invented for the MS model, it can also be made applicable when the variance is inflated.

Thompson [11] uses the VI model for a single outlier in the framework of the restricted (or residual) maximum likelihood estimation, which is known as REML. In contrast to the true maximum likelihood estimation, REML can produce unbiased estimates of variance and covariance parameters and it causes less computational workload. Thompson [11] proposes that the observation with the largest log-likelihood value can be investigated as a possible outlier. Gumedze et al. [12] take up this development and set up a so-called variance shift outlier model (VSOM). Likelihood ratio (LR) and score test statistics are used to identify the outliers. The authors conclude that VSOM gives an objective compromise between including and omitting an observation, where its status as a correct or erroneous observation cannot be adequately resolved. Gumedze [13] review this approach and work out a one-step LR test, which is a computational simplification of the full-step LR test.

In geodesy, the VI model was introduced by Koch and Kargoll [14]. The estimation of the unknown parameters has been established as an iteratively reweighted least squares adjustment. The expectation maximization (EM) algorithm is used to detect the outliers. It is found that the EM algorithm for the VI model is very sensitive to outliers, due to its adaptive estimation, whereas the EM algorithm for the MS model provides the distinction between outliers and good observations. Koch [15] applies the method to fit a surface in three-dimensional space to the Cartesian coordinates of a point cloud obtained from measurements with a laser scanner.

The main goal of this contribution is a detailed derivation of the VI approach in the framework of outlier detection and compare it with the well-established MS model. The performance of both approaches is to be compared in order to derive recommendations for the practical application of outlier detection. This comprises the following objectives:

1. Definition of the generally accepted null model and specification of alternative MS and VI models (Section 2). Multiple outliers are allowed for in both alternative models to keep the models equivalent.
2. True maximization of the likelihood functions of the null and alternative models, not only for the common MS model, but also for the VI model. This means, we do not resort to the REML approach of Thompson [11], Gumedze et al. [12], and Gumedze [13]. This is important for the purpose of an insightful comparison of MS and VI (Section 3).
3. Application of likelihood ratio (LR) test for outlier detection by hypothesis testing and derivation of the test statistics for both the MS and the VI model. For this purpose, a completely new rigorous likelihood ratio test in the VI model is developed and an also completely new comparison with the equivalent test in the MS model is elaborated (Section 3).

4. Comparison of both approaches using the illustrative example of repeated observations, which is worked out in full detail (Section 4).

Section 5 briefly summarises the investigations that were carried out and critically reviews the results. Recommendations for the practical application of outlier detection conclude this paper.

2. Null Model, Mean Shift Model, and Variance Inflation Model

In mathematical statistics, a hypothesis H is a proposed explanation that the probability distribution of the random n-vector y of observations belongs to a certain parametric family W of probability distributions with parameter vector θ, e.g., Teunissen [7],

$$H : y \sim W(\theta), \quad \theta \in \Theta \tag{1}$$

The parameter vector θ might assume values from a set Θ of admissible parameter vectors. A model is then simply the formulation of the relationship between observations y and parameters θ based on H. In geodesy, the standard linear model is based on the hypothesis that the observations follow a normal distribution N, e.g., Koch [16], Teunissen [7], i.e.,

$$H_0 : y \sim N(Ax, \Sigma), \tag{2}$$

with u-vector of functional parameters x and covariance matrix Σ. The latter matrix might contain further stochastic parameters, like a variance factor σ^2, according to

$$\Sigma = \sigma^2 Q, \tag{3}$$

with Q being the known cofactor matrix of y. In this case, θ is the union of x and σ^2. The covariance matrix Σ might eventually contain more stochastic parameters, known as variance components, cf. Koch ([16], p. 225ff). Matrix A is said to be the $n \times u$-matrix of design. The model that is based on H_0 is called the null model.

In outlier detection, we oppose H_0 with one or many alternative hypotheses, most often in the form of a mean shift (MS) hypothesis, e.g., Koch [16], Teunissen [7]

$$H_{MS} : y \sim N(Ax + C\nabla, \Sigma), \quad \nabla \neq 0, \tag{4}$$

where ∇ is a m-vector of additional functional bias parameters and matrix C extends the design. In this case, θ is extended by ∇. This relationship gives rise to the MS model, where the mean of the observations is shifted from Ax to $Ax + C\nabla$ by the effect of gross errors, see Figure 1. $C\nabla$ can be interpreted as accounting for the systematic effect of gross observation errors superposing the effect of normal random observation errors already taken into account by Σ in (2). The great advantage of H_{MS} is that, if the null model is linear or linearized, so is the MS model. The determination of the model parameters is numerically easy and computationally efficient, cf. Lehmann and Lösler [5].

However, there are different possibilities to set up an alternative hypothesis. The most simple one is the variance inflation hypothesis

$$H_{VI} : y \sim N(Ax, \Sigma'), \tag{5}$$

where Σ' is a different covariance matrix, which consists of inflated variances. Σ' can be interpreted as accounting for the joint random effect of normal observation errors in all observations and zero mean gross errors in few outlying observations, see Figure 2.

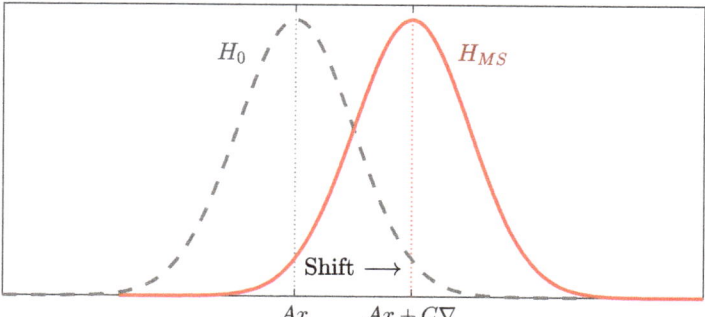

Figure 1. Schematic representation of the MS approach. The null model H_0 is depicted by a dashed grey line and the alternative model H_{MS} is shown as a solid red line.

The VI model might be considered to be more adequate to describe the outlier situation when the act of falsification of the outlying observations is thought of as being a random event, which might not be exactly reproduced in a virtual repetition of the observations. However, even if the VI model might be more adequate to describe the stochastics of the observations, this does not mean that it is possible to estimate parameters or to detect outliers better than with some less adequate model like MS. This point will be investigated below.

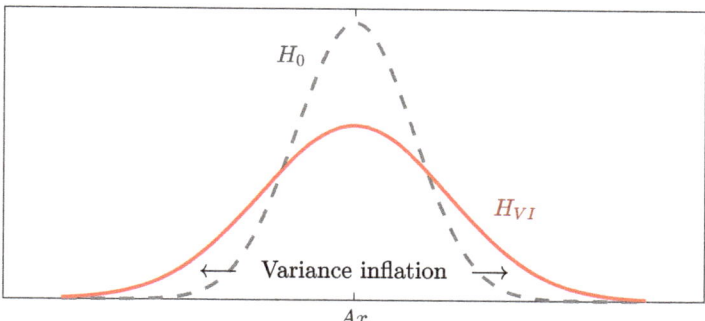

Figure 2. Schematic representation of the VI approach. The null model H_0 is depicted by a dashed grey line and the alternative model H_{VI} is shown as solid red line.

In the following we will only consider the case that y are uncorrelated observations, where both Σ and Σ' are diagonal matrices, such that the hypotheses read

$$H_0 : y \sim N(Ax, \sigma_1^2, \ldots, \sigma_n^2), \tag{6a}$$

$$H_{VI} : y \sim N(Ax, \tau_1 \sigma_1^2, \ldots, \tau_m \sigma_m^2, \sigma_{m+1}^2, \ldots, \sigma_n^2), \quad \tau_1 > 1, \ldots, \tau_m > 1. \tag{6b}$$

Here, H_{VI} accounts for m random zero mean gross errors in the observations $y_1, \ldots, y_m$ modeled by stochastic parameters $\tau_1, \ldots, \tau_m$. In this case, θ is extended by $\tau_1, \ldots, \tau_m$, which will be called variance inflation factors. Thus, $\tau_1, \ldots, \tau_m$ can be interpreted as a special case of extra variance components.

Note that the term variance inflation factors is used differently in multiple linear regression when dealing with multicollinearity, cf. James et al. ([17], p. 101f).

3. Outlier Detection by Hypothesis Tests

Outlier detection can be characterised as a statistical model selection problem. The null model, which describes the expected stochastic properties of the data, is opposed to one or more alternative models, which deviate from such properties. Usually, the decision, whether the null model is rejected in favour of a proper alternative model, is based on hypothesis testing. Multiple testing, consisting of a sequence of testings with one single alternative hypothesis only, is required if there are many possible alternative models. In this study, we first focus on such one single testing only. In the following subsections, the test statistics in the MS model as well as in the VI model are derived.

For the sake of simplicity, the scope of this contribution is restricted to cases, where all of the estimates to be computed are unique, such that all matrices to be inverted are regular.

3.1. Mean Shift Model

In the case of no stochastic parameters, i.e., Σ is known, the optimal test statistic $T_{MS}(y)$ for the test problem H_0 in (2) versus H_{MS} in (4) is well known and yields, cf. Teunissen ([7], p. 76):

$$T_{MS}(y) := \hat{\nabla}^T \Sigma_{\hat{\nabla}}^{-1} \hat{\nabla} \qquad (7)$$
$$= \hat{e}_0^T \Sigma^{-1} C (C^T \Sigma^{-1} \Sigma_{\hat{e}_0}^{-1} \Sigma^{-1})^{-1} C^T \Sigma^{-1} \hat{e}_0,$$

where $\hat{\nabla}$ and $\Sigma_{\hat{\nabla}}$ are the vector of estimated bias parameters in the MS model and its covariance matrix, respectively. Furthermore, $\hat{e}_0$ and $\Sigma_{\hat{e}_0}$ are the vector of estimated residuals $e := Ax - y$ in the null model and its covariance matrix, respectively. This second expression offers the opportunity to perform the test purely based on the estimation in the null model, cf. Teunissen ([7], p. 75), i.e.,

$$\hat{x}_0 = (A^T \Sigma^{-1} A)^{-1} A^T \Sigma^{-1} y, \qquad (8a)$$
$$\hat{e}_0 = (I - A(A^T \Sigma^{-1} A)^{-1} A^T \Sigma^{-1}) y, \qquad (8b)$$
$$\Sigma_{\hat{e}_0} = \Sigma - A(A^T \Sigma^{-1} A)^{-1} A^T. \qquad (8c)$$

Each estimation is such that the likelihood function of the model is maximized. The hat will indicate maximum likelihood estimates below.

The test statistic (7) follows the central or non-central χ^2 distributions

$$T_{MS}|H_0 \sim \chi^2(q, 0), \qquad (9a)$$
$$T_{MS}|H_{MS} \sim \chi^2(q, \lambda), \qquad (9b)$$

where $q = \text{rank}(\Sigma_{\hat{\nabla}})$ is the degree of freedom and $\lambda = \nabla^T \Sigma_{\hat{\nabla}}^{-1} \nabla$ is called the non-centrality parameter, depending on the true but unknown bias parameters ∇, e.g., (Teunissen [7], p. 77).

The test statistic (7) has the remarkable property of being uniformly the most powerful invariant (UMPI). This means, given a probability of type 1 decision error (rejection of H_0 when it is true) α, (7)

- has the least probability of type 2 decision error (failure to reject H_0 when it is false) β (most powerful);
- is independent of ∇ (uniform); and,
- but only for some transformed test problem (invariant).

For the original test problem, no uniformly most powerful (UMP) test exists. For more details see Arnold [18], Kargoll [8].

Lehmann and Voß-Böhme [19] prove that (7) has the property of being a UMPχ^2 test, i.e., a UMP test in the class of all tests with test statistic following a χ^2 distribution. It can be shown that (7) belongs to the class of likelihood ratio (LR) tests, where the test statistic is equivalent to the ratio

$$\frac{\max L_0(x)}{\max L_{MS}(x, \nabla)}. \qquad (10)$$

Here, L_0 and L_{MS} denote the likelihood functions of the null and alternative model, respectively, cf. Teunissen ([7], p. 53), Kargoll [8]. (Two test statistics are said to be equivalent, if they always define the same critical region and, therefore, bring about the same decision. A sufficient condition is that one test statistic is a monotone function of the other. In this case, either both or none exceed their critical value referring to the same α.) The LR test is very common in statistics for the definition of a test statistic, because only a few very simple test problems permit the construction of a UMP test. A justification for this definition is provided by the famous Neyman-Pearson lemma, cf. Neyman and Pearson [20].

3.2. Variance Inflation Model

For the test problem (6a) versus (6b), no UMP test exists. Therefore, we also resort to the LR test here. We start by setting up the likelihood functions of the null and alternative model. For the null model (6a), the likelihood function reads

$$L_0(x) = (2\pi)^{-\frac{n}{2}} \prod_{i=1}^{n} \sigma_i^{-1} \exp\left\{-\frac{\Omega_0}{2}\right\}, \tag{11a}$$

$$\Omega_0 := \sum_{i=1}^{n} \frac{(y_i - a_i x)^2}{\sigma_i^2}, \tag{11b}$$

and for the alternative model (6b) the likelihood function is given by

$$L_{VI}(x, \tau_1, ..., \tau_m) = (2\pi)^{-\frac{n}{2}} \prod_{i=1}^{n} \sigma_i^{-1} \prod_{i=1}^{m} \tau_i^{-\frac{1}{2}} \exp\left\{-\frac{\Omega_{VI}}{2}\right\}, \tag{12a}$$

$$\Omega_{VI} := \sum_{i=1}^{m} \frac{(y_i - a_i x)^2}{\sigma_i^2 \tau_i} + \sum_{i=m+1}^{n} \frac{(y_i - a_i x)^2}{\sigma_i^2}, \tag{12b}$$

where $a_i, i = 1, \ldots, n$ are the row vectors of A. According to (10), the likelihood ratio reads

$$\frac{\max L_0(x)}{\max L_{VI}(x, \tau_1, ..., \tau_m)} = \frac{\max \exp\left\{-\frac{\Omega_0}{2}\right\}}{\max \prod_{i=1}^{m} \tau_i^{-\frac{1}{2}} \exp\left\{-\frac{\Omega_{VI}}{2}\right\}}. \tag{13}$$

Equivalently, we might use the double negative logarithm of the likelihood ratio as test statistic, because it brings about the same decision as the likelihood ratio itself, i.e.,

$$T_{VI} := -2\log \frac{\max \exp\left\{-\frac{\Omega_0}{2}\right\}}{\max \prod_{i=1}^{m} \tau_i^{-\frac{1}{2}} \exp\left\{-\frac{\Omega_{VI}}{2}\right\}}$$

$$= \min \Omega_0 - \min\left\{\Omega_{VI} + \sum_{i=1}^{m} \log \tau_i\right\}. \tag{14}$$

The first minimization result is the well-known least squares solution (8). The second minimization must be performed not only with respect to x, but also with respect to the unknown variance inflation factors $\tau_1, \ldots, \tau_m$. The latter yield the necessary conditions

$$\hat{\tau}_i = \frac{(y_i - a_i \hat{x}_{VI})^2}{\sigma_i^2} = \frac{\hat{e}_{VI,i}^2}{\sigma_i^2}, \quad i = 1, \ldots, m. \tag{15}$$

This means that $\tau_1, \ldots, \tau_m$ are estimated, such that the first m residuals in the VI model $\hat{e}_{VI,i}$ equal in magnitude their inflated standard deviations $\sigma_i\sqrt{\hat{\tau}_i}$, and the subtrahend in (14) is obtained by

$$\min\left\{\Omega_{VI} + \sum_{i=1}^{m}\log\tau_i\right\} = \min\left\{m + \sum_{i=1}^{m}\log\frac{(y_i - a_i x_{VI})^2}{\sigma_i^2} + \sum_{i=m+1}^{n}\frac{(y_i - a_i x_{VI})^2}{\sigma_i^2}\right\} \quad (16)$$

$$= \min\left\{m + \sum_{i=1}^{m}\log\frac{e_{VI,i}^2}{\sigma_i^2} + \sum_{i=m+1}^{n}\frac{e_{VI,i}^2}{\sigma_i^2}\right\}.$$

In the latter expression the minimum is to be found only with respect to the free parameter vector x_{VI}. This expression differs from $\min\Omega_0$ essentially by the logarithm of the first m normalized residuals. This means that those summands are down-weighted, whenever the residuals $\hat{e}_{VI,i}$ are larger in magnitude than their non-inflated standard deviations σ_i. The necessary conditions for x_{VI} are obtained by nullifying the first derivatives of (16) and read

$$0 = \sum_{i=1}^{m}\frac{a_{ij}}{y_i - a_i \hat{x}_{VI}} + \sum_{i=m+1}^{n}\frac{y_i - a_i \hat{x}_{VI}}{\sigma_i^2}a_{ij}, \quad j = 1,\ldots,u, \quad (17)$$

where a_{ij} denotes the j-th element of a_i.

This system of equations can be rewritten as a system of polynomials of degree $m+1$ in the parameters $\hat{x}_{VI,i}$. In general, the solution for $\hat{x}_{VI}$ must be executed by a numerical procedure. This extra effort is certainly a disadvantage of the VI model.

Another disadvantage is that (14) does not follow a well known probability distribution, which complicates the computation of the critical value, being the quantile of this distribution. Such a computation is best performed by Monte Carlo integration, according to Lehmann [21].

Note that the likelihood function L_{VI} in (12) has poles at $\tau_i = 0$, $i = 1,\ldots,m$. These solutions must be excluded from consideration, because they belong to minima of (12) or equivalently to maxima of (16). (Note that $\log\tau_i$ is dominated by $1/\tau_i$ at $\tau_i \to 0$).

A special issue in the VI model is what to do if $\max\hat{\tau}_i \leq 1$ is found in (15). In this case, the variance is not inflated, such that H_0 must not be rejected in favour of H_{VI}, see (6b). However, it might happen that, nonetheless, T_{VI} in (14) exceeds its critical value, especially if α is large. In order to prevent this behaviour, we modify (14) by

$$T_{VI}(y) := \begin{cases} 0 & \text{if } \max\hat{\tau}_i \leq 1, \\ \min\Omega_0 - \min\{\Omega_{VI} + \sum_{i=1}^{m}\log\tau_i\} & \text{otherwise.} \end{cases} \quad (18)$$

If H_0 is true, then there is a small probability that $\max\hat{\tau}_i > 1$ and, consequently, $T_{VI} > 0$ arises, i.e.,

$$\Pr(T_{VI} > 0 | H_0) =: \alpha_{\max}. \quad (19)$$

We see that a type 1 error cannot be required more probable than this $\alpha_{\max}$, i.e., contrary to the MS model, there is an upper limit for the choice of α.

Even more a problem is what to do, if $\min\hat{\tau}_i < 1 < \max\hat{\tau}_i$ is found in (15). Our argument is that, in this case, H_0 should be rejected, but possibly not in favour of H_{VI} in (6b). A more suitable alternative hypothesis should be found in the framework of a multiple test.

4. Repeated Observations

There is one case, which permits an analytical treatment, even of the VI model, i.e., when one scalar parameter x is observed directly n times, such that we obtain $A = (1,\ldots,1)^T =: \mathbf{1}$. By transformation of the observations, also all other models with $u = 1$ can be mapped to this case. For compact notation, we define the weighted means of all observations and of only the last $n - m$ inlying observations, i.e.,

$$w := \frac{\sum_{i=1}^n y_i \sigma_i^{-2}}{\sum_{i=1}^n \sigma_i^{-2}}, \tag{20a}$$

$$W := \frac{\sum_{i=m+1}^n y_i \sigma_i^{-2}}{\sum_{i=m+1}^n \sigma_i^{-2}}. \tag{20b}$$

By covariance propagation, the related variances of those expressions are obtained, i.e.,

$$\sigma_w^2 = \frac{1}{\sum_{i=1}^n \sigma_i^{-2}}, \tag{21a}$$

$$\sigma_W^2 = \frac{1}{\sum_{i=m+1}^n \sigma_i^{-2}}. \tag{21b}$$

Having the following useful identities

$$\frac{w}{\sigma_w^2} = \frac{W}{\sigma_W^2} + \sum_{i=1}^m \frac{y_i}{\sigma_i^2}, \tag{22a}$$

$$\frac{1}{\sigma_w^2} = \frac{1}{\sigma_W^2} + \sum_{i=1}^m \frac{1}{\sigma_i^2}, \tag{22b}$$

$$w - W = \sigma_w^2 \left(\frac{W}{\sigma_W^2} - \frac{W}{\sigma_w^2} \right) + \sigma_w^2 \sum_{i=1}^m \frac{y_i}{\sigma_i^2}$$

$$= \sigma_w^2 \sum_{i=1}^m \frac{y_i - W}{\sigma_i^2}, \tag{22c}$$

the estimates in the null model (8) can be expressed as

$$\hat{x}_0 = w, \tag{23a}$$

$$\hat{e}_0 = y - \mathbf{1}w, \tag{23b}$$

$$\Sigma_{\hat{e}_0} = \Sigma - \mathbf{1}\mathbf{1}^T \sigma_w^2, \tag{23c}$$

and the minimum of the sum of the squared residuals is

$$\min \Omega_0 = \sum_{i=1}^n \frac{(y_i - w)^2}{\sigma_i^2}. \tag{23d}$$

4.1. Mean Shift Model

In the MS model, the first m observations are falsified by bias parameters $\nabla_1, \ldots, \nabla_m$. Matrix

$$C = \begin{pmatrix} I \\ 0 \end{pmatrix} \tag{24}$$

in (4) is a block matrix of the $m \times m$ identity matrix and a $(n - m) \times m$ null matrix. Maximizing the likelihood function yields the estimated parameters and residuals, i.e.,

$$\hat{x}_{MS} = W, \tag{25a}$$

$$\hat{e}_{MS} = y - \mathbf{1}W, \tag{25b}$$

respectively, as well as the estimated bias parameters and their related covariance matrix, i.e.,

$$\hat{\nabla} = \begin{pmatrix} y_1 - W \\ \vdots \\ y_m - W \end{pmatrix} = \begin{pmatrix} \hat{e}_{MS,1} \\ \vdots \\ \hat{e}_{MS,m} \end{pmatrix}, \tag{25c}$$

$$\Sigma_{\hat{\nabla}} = \begin{pmatrix} \sigma_1^2 + \sigma_W^2 & \sigma_W^2 & \cdots & \sigma_W^2 \\ \sigma_W^2 & \sigma_2^2 + \sigma_W^2 & \cdots & \sigma_W^2 \\ \vdots & \vdots & \ddots & \vdots \\ \sigma_W^2 & \sigma_W^2 & \cdots & \sigma_m^2 + \sigma_W^2 \end{pmatrix}, \tag{25d}$$

respectively. Note that (25d) is obtained by covariance propagation that was applied to (25c). By applying the Sherman—Morrison formula, cf. Sherman and Morrison [22], the inverse matrix of $\Sigma_{\hat{\nabla}}$ is obtained,

$$\Sigma_{\hat{\nabla}}^{-1} = \left[\begin{pmatrix} \sigma_1^2 & 0 & \cdots & 0 \\ 0 & \sigma_2^2 & \cdots & 0 \\ \vdots & \vdots & \ddots & \vdots \\ 0 & 0 & \cdots & \sigma_m^2 \end{pmatrix} + \sigma_W^2 \mathbf{1}\mathbf{1}^T \right]^{-1}$$

$$= \begin{pmatrix} \sigma_1^{-2} & 0 & \cdots & 0 \\ 0 & \sigma_2^{-2} & \cdots & 0 \\ \vdots & \vdots & \ddots & \vdots \\ 0 & 0 & \cdots & \sigma_m^{-2} \end{pmatrix} - \frac{\sigma_W^2}{1 + \sigma_W^2 \sum_{i=1}^m \sigma_i^{-2}} \begin{pmatrix} \sigma_1^{-2} \\ \sigma_2^{-2} \\ \vdots \\ \sigma_m^{-2} \end{pmatrix} \begin{pmatrix} \sigma_1^{-2} \\ \sigma_2^{-2} \\ \vdots \\ \sigma_m^{-2} \end{pmatrix}^T$$

$$= \begin{pmatrix} \sigma_1^{-2} & 0 & \cdots & 0 \\ 0 & \sigma_2^{-2} & \cdots & 0 \\ \vdots & \vdots & \ddots & \vdots \\ 0 & 0 & \cdots & \sigma_m^{-2} \end{pmatrix} - \sigma_w^2 \begin{pmatrix} \sigma_1^{-4} & \sigma_1^{-2}\sigma_2^{-2} & \cdots & \sigma_1^{-2}\sigma_m^{-2} \\ \sigma_1^{-2}\sigma_2^{-2} & \sigma_2^{-4} & \cdots & \sigma_2^{-2}\sigma_m^{-2} \\ \vdots & \vdots & \ddots & \vdots \\ \sigma_1^{-2}\sigma_m^{-2} & \sigma_2^{-2}\sigma_m^{-2} & \cdots & \sigma_m^{-4} \end{pmatrix}, \tag{26}$$

and the test statistic (7) in the MS model becomes

$$T_{MS}(y) = \sum_{i=1}^m \frac{\hat{e}_{MS,i}^2}{\sigma_i^2} - \sigma_w^2 \sum_{i=1}^m \sum_{j=1}^m \frac{\hat{e}_{MS,i}\hat{e}_{MS,j}}{\sigma_i^2 \sigma_j^2}. \tag{27}$$

According to (9), the distributions of the null model and the alternative model are given by

$$T_{MS}|H_0 \sim \chi^2(m, 0), \tag{28a}$$

$$T_{MS}|H_{MS} \sim \chi^2(m, \lambda), \tag{28b}$$

respectively, where the non-centrality parameter reads

$$\lambda = \sum_{i=1}^m \frac{\nabla_i^2}{\sigma_i^2} - \sigma_w^2 \sum_{i=1}^m \sum_{j=1}^m \frac{\nabla_i \nabla_j}{\sigma_i^2 \sigma_j^2}. \tag{29}$$

For the special cases of $m = 1$ and $m = 2$ extra bias parameters, as well as the case of independent and identically distributed random observation errors, the related test statistics (27) are given by

Case $m = 1$:
$$T_{MS}(y) = \frac{\hat{e}_{MS,1}^2}{\sigma_1^2 + \sigma_W^2}, \tag{30}$$

Case $m = 2$:
$$T_{MS}(y) = \hat{e}_{MS,1}^2 \frac{\sigma_1^2 - \sigma_w^2}{\sigma_1^4} + \hat{e}_{MS,2}^2 \frac{\sigma_2^2 - \sigma_w^2}{\sigma_2^4} - 2\sigma_w^2 \frac{\hat{e}_{MS,1}\hat{e}_{MS,2}}{\sigma_1^2 \sigma_2^2}, \tag{31}$$

Case $\sigma_1 = \sigma_2 = \cdots = \sigma_m =: \sigma$:
$$T_{MS}(y) = \frac{1}{\sigma^2} \left(\sum_{i=1}^{m} \hat{e}_{MS,i}^2 - \frac{1}{n} \sum_{i=1}^{m} \sum_{j=1}^{m} \hat{e}_{MS,i} \hat{e}_{MS,j} \right). \tag{32}$$

In the case of $\sigma_w \ll \min \sigma_i$, which often arises when $m \ll n$, the test statistic tends to
$$T_{MS}(y) \to \sum_{i=1}^{m} \frac{\hat{e}_{MS,i}^2}{\sigma_i^2}. \tag{33}$$

4.2. Variance Inflation Model—General Considerations

In the VI model, the first m observations are falsified by variance inflation factors $\tau_1, \ldots, \tau_m$. The necessary condition (17) reads
$$0 = \sum_{i=1}^{m} \frac{1}{y_i - \hat{x}_{VI}} + \sum_{i=m+1}^{n} \frac{y_i - \hat{x}_{VI}}{\sigma_i^2}. \tag{34}$$

Using (20b) and (21b), this can be rewritten to
$$\hat{x}_{VI} - W = \sum_{i=1}^{m} \frac{\sigma_W^2}{y_i - \hat{x}_{VI}}. \tag{35}$$

This solution $\hat{x}_{VI}$ is obtained as the real root of a polynomial of degree $m + 1$, which might have, at most $m + 1$, real solutions. In the model, it is easy to exclude the case that $y_i = y_j, i \neq j$, because they are either both outliers or both good observations. They should be merged into one observation. Let us index the observations, as follows: $y_1 < y_2 < \cdots < y_m$. We see that

- in the interval $-\infty \ldots y_1$ of $\hat{x}_{VI}$ the right hand side of (35) goes from 0 to $+\infty$,
- in each interval $y_{i-1} \ldots y_i$ it goes from $-\infty$ to $+\infty$, and
- in the interval $y_m \cdots +\infty$ it goes from $-\infty$ to 0.
- The left hand side of (35) is a straight line.

Therefore, (35) has always at least one real solution $\hat{x}_{VI}$ in each interval $y_{i-1} \ldots y_i$, where one of them must be a maximum of (16), because (16) goes from $-\infty$ up to some maximum and then down again to $-\infty$ in this interval. Besides these $m - 1$ uninteresting solutions, (35) can have no more or two more real solutions, except in rare cases, where it might have one more real solution. If $W < y_1$, then there are no solutions above y_m. If $W > y_m$, then there are no solutions below y_1.

From these considerations it becomes clear that (35) can have, at most, one solution that is a minimum of (16), see also Figure 3.

The second-order sufficient condition for a strict local minimum of (16) is that the Hessian matrix H of (16),

$$
H(x_{VI}, \tau_1, \ldots, \tau_m) = 2 \begin{pmatrix} \frac{1}{\sigma_W^2} + \sum_{i=1}^m \frac{1}{\tau_i^2 \sigma_i^2} & \frac{y_1 - x_{VI}}{\tau_1^2 \sigma_1^2} & \cdots & \frac{y_m - x_{VI}}{\tau_m^2 \sigma_m^2} \\ \frac{y_1 - x_{VI}}{\tau_1^2 \sigma_1^2} & \frac{(y_1 - x_{VI})^2}{\tau_1^3 \sigma_1^2} - \frac{1}{2\tau_1^2} & \cdots & 0 \\ \vdots & \vdots & \ddots & \vdots \\ \frac{y_m - x_{VI}}{\tau_m^2 \sigma_m^2} & 0 & \cdots & \frac{(y_m - x_{VI})^2}{\tau_m^3 \sigma_m^2} - \frac{1}{2\tau_m^2} \end{pmatrix}, \quad (36)
$$

must be positive-definite at $\hat{x}_{VI}, \hat{\tau}_1, \ldots, \hat{\tau}_m$, cf. Nocedal and Wright ([23] p.16), i.e.,

$$
H(\hat{x}_{VI}, \hat{\tau}_1, \ldots, \hat{\tau}_m) = 2 \begin{pmatrix} \frac{1}{\sigma_W^2} + \sum_{i=1}^m \frac{1}{\hat{\tau}_i \sigma_i^2} & \frac{y_1 - \hat{x}_{VI}}{\hat{\tau}_1^2 \sigma_1^2} & \cdots & \frac{y_m - \hat{x}_{VI}}{\hat{\tau}_m^2 \sigma_m^2} \\ \frac{y_1 - \hat{x}_{VI}}{\hat{\tau}_1^2 \sigma_1^2} & \frac{1}{2\hat{\tau}_1^2} & \cdots & 0 \\ \vdots & \vdots & \ddots & \vdots \\ \frac{y_m - \hat{x}_{VI}}{\hat{\tau}_m^2 \sigma_m^2} & 0 & \cdots & \frac{1}{2\hat{\tau}_m^2} \end{pmatrix}. \quad (37)
$$

must be a positive definite matrix. A practical test for positive definiteness that does not require explicit calculation of the eigenvalues is the principal minor test, also known as Sylvester's criterion. The k-th leading principal minor is the determinant that is formed by deleting the last $n - k$ rows and columns of the matrix. A necessary and sufficient condition that a symmetric $n \times n$ matrix is positive definite is that all n leading principal minors are positive, cf. Prussing [24], Gilbert [25]. Invoking Schur's determinant identity, i.e.,

$$
\det \begin{pmatrix} A & B \\ C & D \end{pmatrix} = \det(D) \det(A - BD^{-1}C) \quad (38)
$$

and in combination with (15), we see that the k-th leading principal minor of H in (37) is

$$
\prod_{i=1}^k \left(\frac{1}{2\hat{\tau}_k^2} \right) \left(\frac{1}{\sigma_W^2} + \sum_{i=1}^m \frac{1}{\hat{\tau}_i \sigma_i^2} - \sum_{i=1}^k \frac{2}{\hat{\tau}_i \sigma_i^2} \right). \quad (39)
$$

To be positive, the second factor must be ensured to be positive for each k. Obviously, if this is true for $k = m$, it is also true for all other k. Therefore, the necessary and sufficient condition for a local minimum of (16) reads

$$
\sum_{i=1}^m \frac{\sigma_W^2}{\hat{\tau}_i \sigma_i^2} = \sum_{i=1}^m \frac{\sigma_W^2}{(y_i - \hat{x}_{VI})^2} < 1. \quad (40)
$$

In other words, if and only if $\hat{x}_{VI}$ is sufficiently far away from all outlying observations, it belongs to a strict local minimum of (16).

Using (23d),(16), the test statistic (18) in the VI model for $\max \hat{\tau}_i > 1$ reads

$$
\begin{aligned}
T_{VI} &= \min \Omega_0 - \min \left\{ \Omega_{VI} + \sum_{i=1}^m \log \tau_i \right\} \quad (41) \\
&= \sum_{i=1}^n \frac{(y_i - w)^2}{\sigma_i^2} - m - \sum_{i=1}^m \log \frac{(y_i - \hat{x}_{VI})^2}{\sigma_i^2} - \sum_{i=m+1}^n \frac{(y_i - \hat{x}_{VI})^2}{\sigma_i^2} \\
&= \sum_{i=1}^n \left(\frac{(y_i - w)^2}{\sigma_i^2} - \frac{(y_i - \hat{x}_{VI})^2}{\sigma_i^2} \right) - m + \sum_{i=1}^m \left(\frac{(y_i - \hat{x}_{VI})^2}{\sigma_i^2} - \log \frac{(y_i - \hat{x}_{VI})^2}{\sigma_i^2} \right) \\
&= (\hat{x}_{VI} - w) \sum_{i=1}^n \frac{2y_i - \hat{x}_{VI} - w}{\sigma_i^2} - m + \sum_{i=1}^m (\hat{\tau}_i - \log \hat{\tau}_i) \\
&= -\frac{(\hat{x}_{VI} - w)^2}{\sigma_w^2} - m + \sum_{i=1}^m (\hat{\tau}_i - \log \hat{\tau}_i).
\end{aligned}
$$

Tracing back the flow sheet of computations, it becomes clear that T_{VI} depends on the observations only through $y_1, \ldots, y_m$ and w or equivalently through $y_1, \ldots, y_m$ and W. These $m + 1$ quantities represent a so-called "sufficient statistic" for the outlier test.

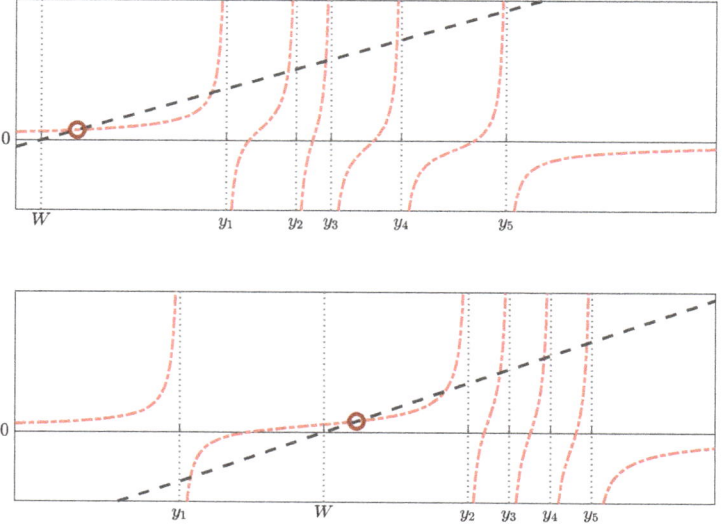

Figure 3. Illustration of two exemplary cases of the solution of (35). Whereas the dash-dot red styled lines indicate the right hand side of the function, the black colored dashed line depicts the left hand side of (35). A dark red circle symbolises the desired solution out of five.

An interesting result is obtained, when we consider the case $\sigma_W \to 0$, which occurs if the $n - m$ good observations contain much more information than the m suspected outliers. In this case, (35) can be rewritten as

$$(\hat{x}_{VI} - W) \prod_{i=1}^{m} (y_i - \hat{x}_{VI}) = \sigma_W^2 P(y_1, \ldots, y_n, \hat{x}_{VI}). \tag{42}$$

where P is some polynomial. If the right hand side goes to zero, at least one factor of the left hand side must also go to zero. As $\sigma_W \to 0$, we obtain $m + 1$ solutions for $\hat{x}_{VI}$, approaching $W, y_1, \ldots, y_m$. The first solution can be a valid VI solution, the others are invalid as $\hat{t}_i \to 0$. Note that we have $\hat{x}_{VI} \to \hat{x}_{MS}$ in this case, and also $\hat{e}_{VI,i} \to \hat{e}_{MS,i}$ for all $i = 1, \ldots, m$. Having in mind (22c), we see that (41) becomes

$$T_{VI} \to -\frac{(W-w)^2}{\sigma_w^2} - m + \sum_{i=1}^{m} (\hat{t}_i - \log \hat{t}_i) \tag{43}$$

$$= \sigma_w^2 \left(\sum_{i=1}^{m} \frac{y_i - W}{\sigma_i^2} \right)^2 - m + \sum_{i=1}^{m} (\hat{t}_i - \log \hat{t}_i) \tag{44}$$

and also noting that with (22b) we find $\sigma_w \to 0$, such that

$$T_{VI} \to -m + \sum_{i=1}^{m} (\hat{\tau}_i - \log \hat{\tau}_i) \tag{45}$$

$$= -m + \sum_{i=1}^{m} \frac{\hat{e}_{MS,i}^2}{\sigma_i^2} - \log \frac{\hat{e}_{MS,i}^2}{\sigma_i^2}. \tag{46}$$

When comparing this to the equivalent result in the MS model (33), we see that T_{VI} and T_{MS} are equivalent test statistics under the sufficient condition $\min \hat{\tau}_i > 1$, because $\tau - \log \tau$ is a monotonic function for $\tau > 1$. This means that, in this case, the decision on H_0 is the same, both in the MS and in the VI model. However, $\max \hat{\tau}_i > 1$ might not be sufficient for this property.

4.3. Variance Inflation Model—Test for One Outlier

In the case $m = 1$ (35) reads

$$\hat{x}_{VI} = \frac{\sigma_W^2}{y_1 - \hat{x}_{VI}} + W \tag{47}$$

Rewriting this to a quadratic equation yields up to two solutions, i.e.,

$$\hat{x}_{VI} = \frac{y_1 + W}{2} \pm \sqrt{\frac{(y_1 - W)^2}{4} - \sigma_W^2}. \tag{48}$$

With (15), we find

$$\hat{\tau}_1 = \frac{1}{\sigma_1^2}\left(\frac{y_1 - W}{2} \pm \sqrt{\frac{(y_1-W)^2}{4} - \sigma_W^2}\right)^2 = \frac{1}{\sigma_1^2}\left(\frac{\hat{e}_{MS,1}}{2} \pm \sqrt{\frac{\hat{e}_{MS,1}^2}{4} - \sigma_W^2}\right)^2. \tag{49}$$

For a solution to exist at all, we must have $|\hat{e}_{MS,1}| \geq 2\sigma_W$. This means that y_1 must be sufficiently outlying, otherwise H_0 is to be accepted.

The condition for a strict local maximum (40) reads here

$$\sigma_W^2 < \hat{\tau}_1 \sigma_1^2 = \left(\frac{\hat{e}_{MS,1}}{2} \pm \sqrt{\frac{\hat{e}_{MS,1}^2}{4} - \sigma_W^2}\right)^2. \tag{50}$$

For the sign of the square root equal to the sign of $\hat{e}_{MS,1}$, this inequality is trivially fulfilled. For the opposite sign, we require

$$\sigma_W < \frac{|\hat{e}_{MS,1}|}{2} - \sqrt{\frac{\hat{e}_{MS,1}^2}{4} - \sigma_W^2}. \tag{51}$$

Rewriting this expression yields

$$\sqrt{\frac{\hat{e}_{MS,1}^2}{4} - \sigma_W^2} < \frac{|\hat{e}_{MS,1}|}{2} - \sigma_W \tag{52}$$

and squaring both sides, which can be done because they are both positive, we readily arrive at $|\hat{e}_{MS,1}| < 2\sigma_W$, which is the case that no solution exists. Therefore, we have exactly one minimum of (16), i.e.,

$$\hat{x}_{VI} = \frac{y_1 + W}{2} - \text{sign}(y_1 - W)\sqrt{\frac{(y_1 - W)^2}{4} - \sigma_W^2} \qquad (53a)$$

$$= \frac{y_1 + W}{2} - \text{sign}(\hat{e}_{MS,1})\sqrt{\frac{\hat{e}_{MS,1}^2}{4} - \sigma_W^2},$$

$$\hat{\tau}_1 = \frac{1}{\sigma_1^2}\left(\frac{\hat{e}_{MS,1}}{2} + \text{sign}(\hat{e}_{MS,1})\sqrt{\frac{\hat{e}_{MS,1}^2}{4} - \sigma_W^2}\right)^2, \qquad (53b)$$

and the test statistic (18) becomes

$$T_{VI} = \begin{cases} 0 & \text{if } \hat{\tau}_1 \leq 1, \\ -\frac{(\hat{x}_{VI} - w)^2}{\sigma_w^2} - 1 + \hat{\tau}_1 - \log \hat{\tau}_1 & \text{otherwise.} \end{cases} \qquad (54)$$

The condition $\hat{\tau}_1 > 1$ is equivalent to

$$\sqrt{\frac{\hat{e}_{MS,1}^2}{4} - \sigma_W^2} > \sigma_1 - \frac{|\hat{e}_{MS,1}|}{2}, \qquad (55)$$

which is trivially fulfilled, if the right hand side is negative. If it is non-negative, both sides can be squared and rearranged to

$$|\hat{e}_{MS,1}| > \frac{\sigma_1^2 + \sigma_W^2}{\sigma_1}. \qquad (56)$$

Since this condition also covers the case that $|\hat{e}_{MS,1}| > 2\sigma_1$, it can be used exclusively as an equivalent of $\hat{\tau}_1 > 1$.

With (22c) we see that both

$$\hat{x}_{VI} - w = \frac{y_1 - W}{2} - \text{sign}(\hat{e}_{MS,1})\sqrt{\frac{\hat{e}_{MS,1}^2}{4} - \sigma_W^2} - \sigma_w^2 \frac{y_1 - W}{\sigma_1^2} \qquad (57)$$

$$= \hat{e}_{MS,1}\left(\frac{1}{2} - \frac{\sigma_w^2}{\sigma_1^2}\right) - \text{sign}(\hat{e}_{MS,1})\sqrt{\frac{\hat{e}_{MS,1}^2}{4} - \sigma_W^2}$$

as well as $\hat{\tau}_1$ through (53b) depend on the observations only through $\hat{e}_{MS,1}$, and so does T_{VI} in (54). On closer examination, we see that T_{VI} in (54) depends even only on $|\hat{e}_{MS,1}|$. This clearly holds as well for T_{MS} in (30). Therefore, both test statistics are equivalent if T_{VI} can be shown to be a strictly monotone function of T_{MS}.

Figure 4 shows that (54) as a function of $\hat{e}_{MS,1}$ is monotone. A mathematical proof of monotony is given in the Appendix A. Thus, it is also monotone as a function of T_{MS} and even strictly monotone for $\hat{\tau}_1 > 1$, which is the case that we are interested in. Therefore, the MS model and the VI model are fully equivalent for repeated observations to be tested for $m = 1$ outlier.

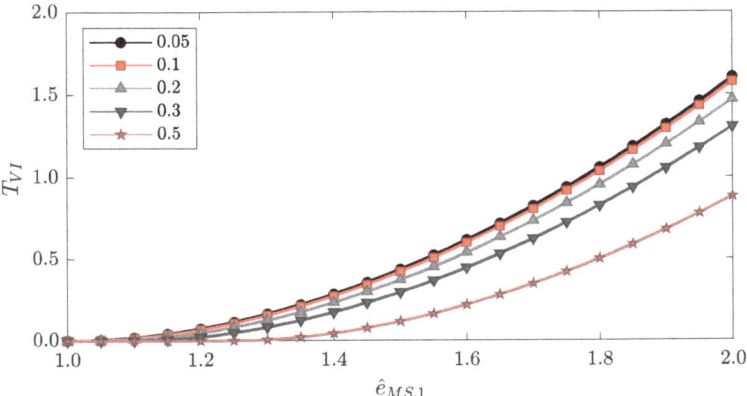

Figure 4. Resulting test statistic T_{VI} of the variance inflation (VI) approach (54) as a function of the residual $\hat{e}_{MS,1}$ for various ratios of σ_w/σ_1.

Finally, we numerically determine the probability distribution of test statistic (54) using Monte Carlo integration by

- defining the ratio σ_w/σ_1,
- generating normally distributed pseudo random numbers for $e_{MS,1}$,
- evaluating (53) and (54), and
- taking the histogram of (54),

using 10^7 pseudo random samples of $\hat{e}_{MS,1}$. In Figure 5, the positive branch of the symmetric probability density function (PDF) is given in logarithmic scale for various ratios of σ_w/σ_1. However, only about 30% of the probability mass is located under this curve, the rest is concentrated at $T_{VI} = 0$, and is not displayed. The quantiles of this distribution determine critical values and also α_{max} in (19). The results are summarized in Table 1.

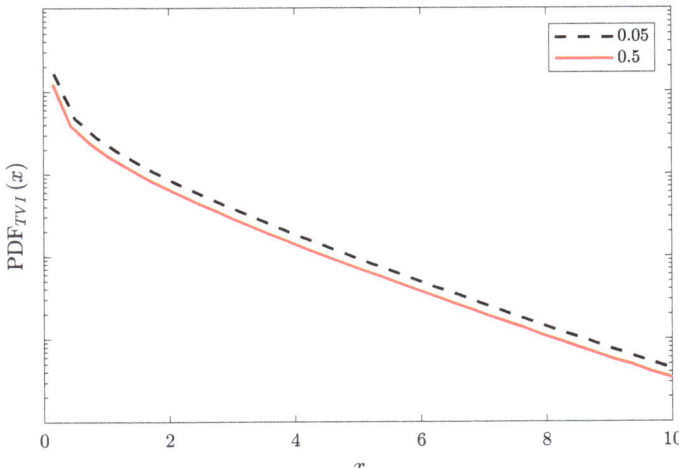

Figure 5. Probability density function of T_{VI} (54) as a function of the residual $\hat{e}_{MS,1}$ under H_0, approximated by MCM. Vertical axis is in logarithmic scale. The strong peaks at 0 for $\hat{\tau}_1 \leq 1$ are not displayed. The dashed black and the solid red curve relates to $\sigma_w/\sigma_1 = 0.05$ and $\sigma_w/\sigma_1 = 0.5$, respectively.

Table 1. Maximum selectable type 1 error probability α_{max} and critical value c_α for T_{VI} in (54) for various ratios of σ_W/σ_1 and $\alpha = 0.05$.

σ_W/σ_1	α_{max}	c_α
0.05	0.317	1.50
0.10	0.315	1.49
0.20	0.308	1.47
0.30	0.297	1.43
0.50	0.264	1.33

4.4. Variance Inflation Model—Test for Two Outliers

In the case $m = 2$ outliers, (35) reads

$$\hat{x}_{VI} - W = \frac{\sigma_W^2}{y_1 - \hat{x}_{VI}} + \frac{\sigma_W^2}{y_2 - \hat{x}_{VI}}. \tag{58}$$

The solution for $\hat{x}_{VI}$ can be expressed in terms of a cubic equation, which permits an analytical solution. One real solution must be in the interval $y_1 \ldots y_2$, but there may be two more solutions

- both below y_1, if $W < y_1$, or
- both above y_2, if $W > y_2$, or
- both between y_1 and y_2, if $y_1 > W > y_2$.

In rare cases, solutions may also coincide. The analytical expressions are very complicated and they do not permit a treatment analogous to the preceding subsection. Therefore, we have to fully rely on numerical methods, which, in our case, is the Monte Carlo method (MCM).

First, we compute the critical values of the test statistic (18)

$$T_{VI} = \begin{cases} 0 & \text{if } \max(\hat{\tau}_1, \hat{\tau}_2) \leq 1, \\ -\frac{(\hat{x}_{VI} - w)^2}{\sigma_w^2} - 2 + \hat{\tau}_1 + \hat{\tau}_2 - \log \hat{\tau}_1 - \log \hat{\tau}_2 & \text{otherwise.} \end{cases} \tag{59}$$

The MCM is preformed, using 10^7 pseudo random samples. We restrict ourselves to the case $\sigma_1 = \sigma_2$. The maximum selectable type 1 error probabilities α_{max} are summarized in Table 2. It is shown that α_{max} is mostly larger than for $m = 1$. The reason is that, more often, we obtain $\hat{\tau}_i > 1$, even if H_0 is true, which makes it easier to define critical values in a meaningful way. Moreover, Table 2 indicates the probabilities that under H_0

- (16) has no local minimum, and if it has, that
- $\max(\hat{\tau}_1, \hat{\tau}_2) \leq 1$
- $\hat{\tau}_1 \leq 1, \hat{\tau}_2 > 1$ or vice versa
- $\min(\hat{\tau}_1, \hat{\tau}_2) > 1$

i.e., none, one, or both variances are inflated. It is shown that, if the good observations contain the majority of the information, a minimum exists, but, contrary to our expectation, the case $\max(\hat{\tau}_1, \hat{\tau}_2) \leq 1$ is not typically the dominating case.

The important result is what happens, if H_0 is false, because variances are truly inflated. The probability that H_0 is rejected is known as the power $1 - \beta$ of the test, where β is the probability of a type 2 decision error. It is computed both with T_{MS} in (31) as well as with T_{VI} in (59). Table 3 provides the results. It is shown that the power of T_{MS} is always better than of T_{VI}. This is unexpected, because T_{MS} is not equivalent to the likelihood ratio of the VI model.

A possible explanation of the low performance of T_{VI} in (59) is that, in many cases, the likelihood function L_{VI} has no local maximum, such that (16) has no local minimum. Even for an extreme variance inflation of $\tau_1 = \tau_2 = 5$ this occurs with remarkable probability of 0.14. Moreover, the probability that $\max(\hat{\tau}_1, \hat{\tau}_2) \leq 1$ is hardly less than that. In both cases, H_0 cannot be rejected.

Table 2. Maximum selectable type 1 error probability α_{max} and critical value c_α for T_{VI} in (59) for various ratios of $\sigma_W/\sigma_1 = \sigma_W/\sigma_2$ and $\alpha = 0.05$ as well as probabilities that (16) has no local minimum or that 0 or 1 or 2 variances are inflated.

				Probabilities for		
				no	0	1	2
$\sigma_W/\sigma_1 = \sigma_W/\sigma_2$	α_{max}	c_α	min	Inflated Variances		
0.01	0.52	4.68	0.03	0.45	0.42	0.10
0.02	0.51	4.36	0.06	0.42	0.41	0.10
0.03	0.50	4.11	0.09	0.40	0.40	0.10
0.05	0.48	3.69	0.15	0.36	0.38	0.10
0.10	0.43	3.00	0.29	0.27	0.33	0.10
0.20	0.33	2.32	0.53	0.14	0.24	0.09
0.30	0.24	1.82	0.69	0.06	0.16	0.09
0.50	0.12	0.97	0.88	0.01	0.05	0.07

Table 3. Test power $1 - \beta_{MS}$ for test using T_{MS} in (31) and test power $1 - \beta_{VI}$ for test using T_{VI} in (59) for various true values of the variance inflation factors τ_1, τ_2 for $\sigma_W = 0.1 \cdot \sigma_1 = 0.1 \cdot \sigma_2$ and $\alpha = 0.05$, as well as probabilities that (16) has no local minimum or that 0 or one or two variances are inflated.

		Test Power		Probabilities for		
				no	0	1	2
τ_1	τ_2	$1 - \beta_{MS}$	$1 - \beta_{VI}$	min	Inflated Variances		
1.0	1.0	0.05	0.05	0.29	0.27	0.33	0.10
1.2	1.2	0.08	0.08	0.27	0.24	0.35	0.13
1.5	1.5	0.13	0.12	0.24	0.21	0.38	0.17
2.0	2.0	0.22	0.20	0.21	0.17	0.39	0.23
3.0	3.0	0.36	0.32	0.18	0.12	0.39	0.31
5.0	5.0	0.55	0.49	0.14	0.08	0.36	0.42
1.0	1.5	0.09	0.09	0.27	0.24	0.36	0.13
1.0	3.0	0.21	0.19	0.24	0.18	0.41	0.18
2.0	3.0	0.30	0.26	0.19	0.14	0.40	0.27
2.0	5.0	0.40	0.36	0.18	0.11	0.40	0.31

4.5. Outlier Identification

If it is not known which observations are outlier-suspected, a multiple test must be set up. If the critical values are identical in all tests, then we simply have to look for the largest test statistic. This is the case for T_{MS} when considering the same number m of outlier-suspected observations, see (9). If we even consider different numbers m in the same multiple test, we have to apply the p-value approach, cf. Lehmann and Lösler [5].

In the VI model, the requirement of identical critical values is rarely met. It is, in general, not met for repeated observations, not even for $m = 1$, as can be seen in Figure 5. However, in this case, it is no problem, because the test with T_{VI} in (54) is equivalent to T_{MS} in (30), as demonstrated. This also means that the same outlier is identified with both test statistics.

For repeated observations, we find identical critical values only for identical variances of the outlier-suspected observations, such that those observations are fully indistinguishable from each other. For example, for $n = 27$, $m = 2$, $\alpha = 0.05$, and $\sigma_1 = \cdots = \sigma_n$, we find $\sigma_W/\sigma_i = 0.20$ and $c_\alpha = 2.32$ for all 351 pairs of outlier-suspected observations, see Table 2.

We evaluate the identifiability of two outliers in $n = 10$ and $n = 20$ repeated observations with $m = 2$ outliers while using the MCM. In each of the 10^6 repetitions, random observations are generated having equal variances. Two cases are considered. Whereas, in the first case, two variances are inflated by τ according to the VI model, in the second case, two observation values are shifted by ∇ according to the MS model. Using (31) and (59), the test statistics T_{MS} and T_{VI} are computed for all $n(n-1)/2 = 45$ or 190 pairs of observations. If the maximum of the test statistic is attained for the actually modified pair of observations, the test statistic correctly identifies the outliers. Here, we assume that α is large enough for the critical value to be exceeded, but otherwise the results are independent of the choice of α. The success probabilities are given in Table 4.

Table 4. Success probabilities for outlier identification in repeated observations of equal variance σ^2 with two outliers.

	Success Probabilities for $n = 10$		Success Probabilities for $n = 20$	
$\tau_1 = \tau_2$	of T_{MS} in (31)	of T_{VI} in (59)	of T_{MS} in (31)	of T_{VI} in (59)
1.0	0.022	0.021	0.005	0.005
2.0	0.065	0.061	0.028	0.026
3.0	0.108	0.104	0.061	0.057
4.0	0.150	0.144	0.095	0.089
5.0	0.185	0.180	0.128	0.121
6.0	0.218	0.212	0.156	0.151
$\nabla_1 = \nabla_2$	of T_{MS} in (31)	of T_{VI} in (59)	of T_{MS} in (31)	of T_{VI} in (59)
0.0	0.022	0.021	0.005	0.005
1σ	0.081	0.065	0.035	0.028
2σ	0.325	0.286	0.226	0.202
3σ	0.683	0.652	0.606	0.586
4σ	0.912	0.902	0.892	0.886
5σ	0.985	0.984	0.984	0.984

As expected, the success probabilities increase as τ or ∇ gets large. However, for both cases, T_{MS} outperforms T_{VI}. In Figure 6, the ratio r_T of the success probabilities between the VI and the MS approach is depicted, having $n = 10$ repeated observations. If $r_T > 1$, the success rate of T_{VI} is higher than for T_{MS} and vice versa. The ratio is always $r_T < 1$ and tends to 1, as shown in Figure 6. Therefore, the success probability of the MS is higher than for the VI approach, even if the outliers are caused by an inflation of the variances.

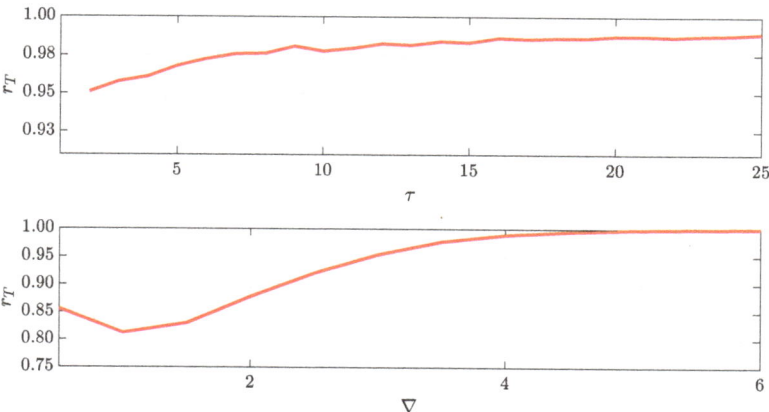

Figure 6. Ratio r_T of the success probabilities between the VI and the MS approach using $n = 10$ repeated observations. The top figure depicts r_T for the case of inflated variances. The bottom figure depicts r_T for the case of shifted mean values.

5. Conclusions

We have studied the detection of outliers in the framework of statistical hypothesis testing. We have investigated two types of alternative hypotheses: the mean shift (MS) hypothesis, where the probability distributions of the outliers are thought of as having a shifted mean, and the variance inflation (VI) model, where they are thought of as having an inflated variance. This corresponds to an outlier-generating process thought of as being deterministic or random, respectively. While the first type of alternative hypothesis is routinely applied in geodesy and in many other disciplines, the second is not. However, even if the VI model might be more adequate to describe the stochastics of the observations, this does not mean that it is possible to estimate parameters or detect outliers better than with some less adequate model, like MS.

The test statistic has been derived by the likelihood ratio of null and alternative hypothesis. This was motivated by the famous Neyman–Pearson lemma, cf. Neyman and Pearson [20], even though that this lemma does not directly apply to this test. Therefore, the performance of the test must be numerically evaluated.

When compared to existing VI approaches, we

- strived for a true (non-restricted) maximization of the likelihood function;
- allowed for multiple outliers;
- fully worked out the case of repeated observations; and,
- computated the corresponding test power by MC method for the first time.

We newly found out that the VI stochastic model has some critical disadvantages:

- the maximization of the likelihood function requires the solution of a system of u polynomial equations of degree $m + 1$, where u is the number of model parameters and m is the number of suspected outliers;
- it is neither guaranteed that the likelihood function actually has such a local maximum, nor that it is unique;
- the maximum might be at a point where some variance is deflated rather than inflated. It is debatable, what the result of the test should be in such a case;
- the critical value of this test must be computed numerically by Monte Carlo integration. This must even be done for each model separately; and,
- there is an upper limit (19) for the choice of the critical value, which may become small in some cases.

For the first time, the application of the VI model has been investigated for the most simple model of repeated observations. It is shown that here the likelihood function admits at most one local maximum, and it does so, if the outliers are strong enough. Moreover, in the limiting case that the suspected outliers represent an almost negligible amount of information, the VI test statistic and the MS test statistic have been demonstrated to be almost equivalent.

For $m = 1$ outlier in the repeated observations, there is even a closed formula (54) for the test statistic, and the existence and uniqueness of a local maximum is equivalent to a simple checkable inequality condition. Additionally, here the VI test statistic and the MS test statistic are equivalent.

In our numerical investigations, we newly found out that for $m > 1$ outliers in the repeated observations the power of the VI test is worse than using the classical MS test statistic. The reason is the lack of a maximum of the likelihood function, even for sizable outliers. Our numerical investigations also show that the identifiability of the outliers is worse for the VI test statistic. This is clearly seen in the case that the outliers are truly caused by shifted means, but also in the other case the identifiability is slightly worse. This means that the correct outliers are more often identified with the MS test statistic.

In the considered cases, we did not find real advantages of the VI model, but this does not prove that they do not exist. As long as such cases are not found, we therefore recommend practically performing outlier detection by the MS model.

Author Contributions: Conceptualization, R.L.; Methodology, R.L.; Software, R.L. and M.L.; Validation, M.L.; Investigation, R.L. and M.L.; Writing—Original Draft Preparation, R.L. and M.L.; Writing—Review & Editing, R.L., M.L., F.N.; Visualization, M.L.; Supervision, F.N. All authors have read and agreed to the published version of the manuscript.

Funding: This research received no external funding.

Conflicts of Interest: The authors declare no conflict of interest.

Appendix A

To prove the monotony of the tails of the function $T_{VI}(e_{MS,1})$ in (54), the extreme values of T_{VI} have to be determined. Since T_{VI} is symmetric, it is sufficient to restrict the proof for positive values of $e_{MS,1}$. The first derivation of T_{VI} is given by

$$T'_{VI}(e_{MS,1}) = -\frac{4\sigma_W^2(\sigma_1^2+1) - e_{MS,1}^2(\sigma_1^2+\sigma_W^2) + e_{MS,1}(\sigma_1^2-\sigma_W^2)\sqrt{e_{MS,1}^2-4\sigma_W^2}}{\sigma_W^2(\sigma_1^2+\sigma_W^2)\sqrt{e_{MS,1}^2-4\sigma_W^2}}. \tag{A1}$$

Setting $T'_{VI}(e_{MS,1}) = 0$ yields two roots

$$e_{\pm} = \pm \frac{(\sigma_1^2+\sigma_W^2)}{\sigma_1}. \tag{A2}$$

Inserting the positive extreme value e_+ into the second derivation of T_{VI}, i.e.,

$$T''_{VI}(e_{MS,1}) = \frac{e_{MS,1}(\sigma_1^2+\sigma_W^2) + \sqrt{e_{MS,1}^2-4\sigma_W^2}(\sigma_W^2-\sigma_1^2)}{\sigma_W^2(\sigma_1^2+\sigma_W^2)\sqrt{e_{MS,1}^2-4\sigma_W^2}}, \tag{A3}$$

identifies e_+ as a minimum value, because $T''_{VI}(e_+)$ is always positive for $\tau_1 > 1$, cf. (55). For that reason, T_{VI} is a monotonically increasing function on the interval $(e_+, +\infty)$. Figure A1 depicts the positive tail of T_{VI} and T'_{VI}, respectively, as well as the minimum e_+.

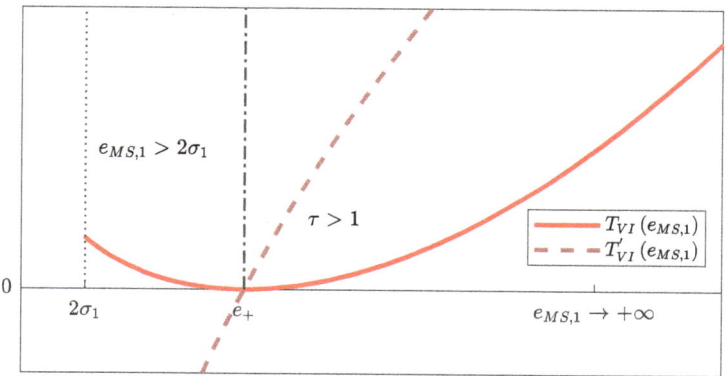

Figure A1. Tail of the function T_{VI} and its first derivation T'_{VI} for positive values of $e_{MS,1}$ as well as the minimum e_+.

References

1. Rofatto, V.F.; Matsuoka, M.T.; Klein, I.; Veronez, M.R.; Bonimani, M.L.; Lehmann, R. A half-century of Baarda's concept of reliability: a review, new perspectives, and applications. *Surv. Rev.* **2018**, 1–17. [CrossRef]
2. Baarda, W. *A Testing Procedure for Use in Geodetic Networks*, 2nd ed.; Netherlands Geodetic Commission, Publication on Geodesy: Delft, The Netherlands, 1968.
3. Teunissen, P.J.G. Distributional theory for the DIA method. *J. Geod.* **2018**, *92*, 59–80. [CrossRef]
4. Lehmann, R. On the formulation of the alternative hypothesis for geodetic outlier detection. *J. Geod.* **2013**, *87*, 373–386. [CrossRef]
5. Lehmann, R.; Lösler, M. Multiple outlier detection: hypothesis tests versus model selection by information criteria. *J. Surv. Eng.* **2016**, *142*, 04016017. [CrossRef]
6. Lehmann, R.; Lösler, M. Congruence analysis of geodetic networks – hypothesis tests versus model selection by information criteria. *J. Appl. Geod.* **2017**, *11*. [CrossRef]
7. Teunissen, P.J.G. *Testing Theory—An Introduction*, 2nd ed.; Series of Mathematical Geodesy and Positioning; VSSD: Delft, The Netherlands, 2006.
8. Kargoll, B. *On the Theory and Application of Model Misspecification Tests in Geodesy*; C 674; German Geodetic Commission: Munich, Germany, 2012.
9. Bhar, L.; Gupta, V. Study of outliers under variance-inflation model in experimental designs. *J. Indian Soc. Agric. Stat.* **2003**, *56*, 142–154.
10. Cook, R.D. Influential observations in linear regression. *J. Am. Stat. Assoc.* **1979**, *74*, 169–174. [CrossRef]
11. Thompson, R. A note on restricted maximum likelihood estimation with an alternative outlier model. *J. R. Stat. Soc. Ser. B (Methodological)* **1985**, *47*, 53–55. [CrossRef]
12. Gumedze, F.N.; Welhamb, S.J.; Gogel, B.J.; Thompson, R. A variance shift model for detection of outliers in the linear mixed model. *Comput. Stat. Data Anal.* **2010**, *54*, 2128–2144. [CrossRef]
13. Gumedze, F.N. Use of likelihood ratio tests to detect outliers under the variance shift outlier model. *J. Appl. Stat.* **2018**. [CrossRef]
14. Koch, K.R.; Kargoll, B. Expectation maximization algorithm for the variance-inflation model by applying the t-distribution. *J. Appl. Geod.* **2013**, *7*, 217–225. [CrossRef]
15. Koch, K.R. Outlier detection for the nonlinear Gauss Helmert model with variance components by the expectation maximization algorithm. *J. Appl. Geod.* **2014**, *8*, 185–194. [CrossRef]
16. Koch, K.R. *Parameter Estimation and Hypothesis Testing in Linear Models*; Springer: Berlin/Heidelberg, Germany, 1999. [CrossRef]
17. James, G.; Witten, D.; Hastie, T.; Tibshirani, R. *An Introduction to Statistical Learning: With Applications in R*; Springer Texts in Statistics; Springer: New York, NY, USA, 2013. [CrossRef]
18. Arnold, S.F. *The Theory of Linear Models and Multivariate Analysis*; Wiley Series in Probability and Statistics; John Wiley & Sons Inc.: New York, NY, USA, 1981.
19. Lehmann, R.; Voß-Böhme, A. On the statistical power of Baarda's outlier test and some alternative. *Surv. Rev.* **2017**, *7*, 68–78. [CrossRef]
20. Neyman, J.; Pearson, E.S. On the problem of the most efficient tests of statistical hypotheses. *Philos. Trans. R. Soc. Lond. Ser. A Contain. Pap. Math. Phys. Character* **1933**, *231*, 289–337. [CrossRef]
21. Lehmann, R. Improved critical values for extreme normalized and studentized residuals in Gauss-Markov models. *J. Geod.* **2012**, *86*, 1137–1146. [CrossRef]
22. Sherman, J.; Morrison, W.J. Adjustment of an inverse matrix corresponding to a change in one element of a given matrix. *Ann. Math. Stat.* **1950**, *21*, 124–127. [CrossRef]
23. Nocedal, J.; Wright, S.J. *Numerical Optimization*, 2nd ed.; Springer: New York, NY, USA, 2006. [CrossRef]
24. Prussing, J.E. The principal minor test for semidefinite matrices. *J. Guid. Control Dyn.* **1986**, *9*, 121–122. [CrossRef]
25. Gilbert, G.T. Positive definite matrices and Sylveste's criterion. *Am. Math. Mon.* **1991**, *98*, 44. [CrossRef]

© 2020 by the authors. Licensee MDPI, Basel, Switzerland. This article is an open access article distributed under the terms and conditions of the Creative Commons Attribution (CC BY) license (http://creativecommons.org/licenses/by/4.0/).

Article

A Generic Approach to Covariance Function Estimation Using ARMA-Models

Till Schubert *, Johannes Korte, Jan Martin Brockmann and Wolf-Dieter Schuh

Institute of Geodesy and Geoinformation, University of Bonn, 53115 Bonn, Germany; korte@geod.uni-bonn.de (J.K.); brockmann@geod.uni-bonn.de (J.M.B.); schuh@uni-bonn.de (W.-D.S.)
* Correspondence: schubert@geod.uni-bonn.de

Received: 27 February 2020; Accepted: 11 April 2020; Published: 15 April 2020

Abstract: Covariance function modeling is an essential part of stochastic methodology. Many processes in geodetic applications have rather complex, often oscillating covariance functions, where it is difficult to find corresponding analytical functions for modeling. This paper aims to give the methodological foundations for an advanced covariance modeling and elaborates a set of generic base functions which can be used for flexible covariance modeling. In particular, we provide a straightforward procedure and guidelines for a generic approach to the fitting of oscillating covariance functions to an empirical sequence of covariances. The underlying methodology is developed based on the well known properties of autoregressive processes in time series. The surprising simplicity of the proposed covariance model is that it corresponds to a finite sum of covariance functions of second-order Gauss–Markov (SOGM) processes. Furthermore, the great benefit is that the method is automated to a great extent and directly results in the appropriate model. A manual decision for a set of components is not required. Notably, the numerical method can be easily extended to ARMA-processes, which results in the same linear system of equations. Although the underlying mathematical methodology is extensively complex, the results can be obtained from a simple and straightforward numerical method.

Keywords: autoregressive processes; ARMA-process; colored noise; continuous process; covariance function; stochastic modeling; time series

---

1. Introduction and Motivation

Covariance functions are an important tool to many stochastic methods in various scientific fields. For instance, in the geodetic community, stochastic prediction or filtering is typically based on the collocation or least squares collocation theory. It is closely related to Wiener–Kolmogorov principle [1,2] as well as to the Best Linear Unbiased Predictor (BLUP) and kriging methods (e.g., [3–5]). Collocation is a general theory, which even allows for a change of functional within the prediction or filtering step, simply propagating the covariance to the changed functional (e.g., [6] (p. 66); [4] (pp. 171–173); [7]). In this context, signal covariance modeling is the crucial part that directly controls the quality of the stochastic prediction or filtering (e.g., [8] (Ch. 5)).

Various references on covariance functions are found in geostatistics [9,10] or atmospheric studies [11–14]. The authors in [4,5,8,15–18] cover the topic of covariance functions in the context of various geodetic applications. The discussion of types of covariance functions and investigations on positive definiteness is covered in the literature in e.g., [13,19–21].

In common practice, covariance modelers tend to use simple analytical models (e.g., exponential and Gauss-type) that can be easily adjusted to the first, i.e., to the short range, empirically derived covariances. Even then, as these functions are nonlinear, appropriate fitting methods are not straightforward. On the other hand, for more complex covariance models, visual diagnostics such as half-value width, correlation length, or curvature are difficult to obtain or even undefined for certain (e.g., oscillatory) covariance models. In addition, fitting procedures to any of these are sparse and lack an automated procedure (cf. [17,22]).

Autoregressive processes (AR-processes) define a causal recursive mechanism on equispaced data. Stochastic processes and autoregressive processes are covered in many textbooks in a quite unified parametrization (e.g., [23–30]). In contrast to this, various parametrizations of Autoregressive Moving Average (ARMA) processes exist in stochastic theory [31–35]. A general connection of these topics, i.e., stochastic processes, covariance functions, and the collocation theory in the context of geodetic usage is given by [5].

In geodesy, autoregressive processes are commonly used in the filter approach as a stochastic model within a least squares adjustment. Decorrelation procedures by digital filters derived from the parametrization of stochastic processes are widely used as they are very flexible and efficient for equidistant data (cf. e.g., [36–42]). Especially for highly correlated data, e.g., observations along a satellite's orbit, advanced stochastic models can be described by stochastic processes. This is shown in the context of gravity field determination from observations of the GOCE mission with the time-wise approach, where flexible stochastic models are iteratively estimated from residuals [43–45].

The fact that an AR-process defines an analytical covariance sequence as well (e.g., [23]) is not well established in geostatistics and geodetic covariance modeling. To counteract this, we relate the covariance function associated with this stochastic processes to the frequently used family of covariance functions of second-order Gauss–Markov (SOGM) processes. Expressions for the covariance functions and fitting methods are aligned to mathematical equations of these stochastic process. Especially in collocation theory, a continuous covariance function is necessary to obtain covariances for arbitrary lags and functionals required for the prediction of multivariate data. However, crucially, the covariance function associated with an AR-process is in fact defined as a discrete sequence (e.g., [26]).

Whereas standard procedures which manually assess decaying exponential functions or oscillating behavior by visual inspection can miss relevant components, for instance high-frequent oscillations in the signal, the proposed method is automated and easily expandable to higher order models in order to fit components which are not obvious at first glance. A thorough modeling of a more complex covariance model does not only result in a better fit of the empirical covariances but results in a beneficial knowledge of the signal process itself.

Within this contribution, we propose an alternative method for the modeling of covariance functions based on autoregressive stochastic processes. It is shown that the derived covariance can be evaluated continuously and corresponds to a sum of well known SOGM-models. To derive the proposed modeling procedure and guidelines, the paper is organized as follows. After a general overview of the related work and the required context of stochastic modeling and collocation theory in Section 2, Section 3 summarizes the widely used SOGM-process for covariance function modeling. Basic characteristics are discussed to create the analogy to the stochastic processes, which is the key of the proposed method. Section 4 introduces the stochastic processes with a special focus on the AR-process and presents the important characteristics, especially the underlying covariance sequence in different representations. The discrete sequence of covariances is continuously interpolated by re-interpretation of the covariance sequence as a difference/differential equation, which has a continuous solution. Based on these findings, the proposed representation can be easily extended to more general ARMA-processes as it is discussed in Section 5. Whereas the previous chapters are based on the consistent definition of the covariance sequences of the processes, Section 6 shows how the derived equations and relations can be used to model covariance functions from empirically estimated covariances in a flexible and numerically simple way. In Section 7, the proposed method is applied to a one-dimensional time series which often serves as an example application for covariance modeling. The numerical example highlights the flexibility and the advantage of the generic procedure. This is followed by concluding remarks in Section 8.

2. Least Squares Collocation

In stochastic theory, a measurement model $\mathcal{L}$ is commonly seen as a stochastic process which consists of a deterministic trend $A\xi$, a random signal $\mathcal{S}$, and a white noise component $\mathcal{N}$ (cf. e.g., [3] (Ch. 2); [5] (Ch. 3.2))

$$\mathcal{L} = A\xi + \mathcal{S} + \mathcal{N}. \tag{1}$$

Whereas the signal part $\mathcal{S}$ is characterized by a covariance stationary stochastic process, the noise $\mathcal{N}$ is usually assumed as a simple white noise process with uncorrelated components. Autocovariance functions $\gamma(|\tau|)$ or discrete sequences $\{\gamma_{|j|}\}_{\Delta t}$ are models to describe the stochastic behavior of the random variables, i.e., generally $\gamma^{\mathcal{S}}(|\tau|)$ and $\gamma^{\mathcal{N}}(|\tau|)$, and are required to be positive semi-definite ([28] (Proposition 1.5.1, p. 26)). In case covariances are given as a discrete sequence $\{\gamma_{|j|}\}_{\Delta t}$, they are defined at discrete lags $h = j\Delta t$ with sampling interval Δt and $j \in \mathbb{N}_0$. In general, autocovariance functions or sequences are functions of a non-negative distance, here τ or h for the continuous and discrete case, respectively. Thus, covariance functions are often denoted by the absolute distance $|\tau|$ and $|h|$. Here, we introduce the conditions $\tau \geq 0$ and $h \geq 0$ in order to omit the vertical bars.

The term Least Squares Collocation, introduced in geodesy by [3,6,46], represents the separability problem within the remove–restore technique, where a deterministic estimated trend component $A\tilde{x}$ is subtracted from the measurements ℓ and the remainder $\widetilde{\Delta\ell} = \ell - A\tilde{x}$ is interpreted as a special realization of the stochastic process $\widetilde{\Delta\mathcal{L}}$. In the trend estimation step

$$\tilde{x} = \left(A^T(\Sigma_{\mathcal{SS}} + \Sigma_{\mathcal{NN}})^{-1} A\right)^{-1} A^T(\Sigma_{\mathcal{SS}} + \Sigma_{\mathcal{NN}})^{-1}\ell, \tag{2}$$

the optimal parameter vector $\tilde{x}$ is computed from the measurements ℓ as the best linear unbiased estimator (BLUE) for the *true* trend $A\xi$. The collocation step follows as the best linear unbiased predictor (BLUP) of the stochastic signal $\tilde{s}'$ at arbitrary points

$$\tilde{s}' = \Sigma_{\mathcal{S'S}}(\Sigma_{\mathcal{SS}} + \Sigma_{\mathcal{NN}})^{-1}\widetilde{\Delta\ell} \tag{3}$$

or as a filter process at the measured points

$$\tilde{s} = \Sigma_{\mathcal{SS}}(\Sigma_{\mathcal{SS}} + \Sigma_{\mathcal{NN}})^{-1}\widetilde{\Delta\ell} \tag{4}$$

([3] (Ch. 2); [5] (Ch. 3)). The variance/covariance matrices $\Sigma_{\mathcal{SS}}$ reflect the stochastic behavior of the random signal $\mathcal{S}$. The coefficients are derived by the evaluation of the covariance function $\gamma^{\mathcal{S}}(\tau)$, where τ represents the lags or distances between the corresponding measurement points. $\Sigma_{\mathcal{NN}}$ denotes the variances/covariances of the random noise which is often modeled as independent and identically distributed random process, $\gamma^{\mathcal{N}}(\tau) = \delta_{0,\tau}\sigma_{\mathcal{N}}^2$ with $\delta_{i,j}$ being the Kronecker delta. $\sigma_{\mathcal{N}}^2$ denotes the variance of the noise such that the covariance matrix reads $\Sigma_{\mathcal{NN}} = \mathbb{1}\sigma_{\mathcal{N}}^2$. $\Sigma_{\mathcal{S'S}}$ is filled row-wise with the covariances of the signal between the prediction points and the measured points.

The *true* covariance functions $\gamma^{\mathcal{S}}(\tau)$ and $\gamma^{\mathcal{N}}(\tau)$ are unknown and often have to be estimated, i.e., $g^{\mathcal{S}}(\tau)$ and $g^{\mathcal{N}}(\tau)$, directly from the trend reduced measurements $\Delta\ell$ by an estimation procedure using the empirical covariance sequences $\{\tilde{g}_j^{\Delta\mathcal{L}}\}_{\Delta t}$. The estimated noise variance $\tilde{s}_{\mathcal{N}}^2$ can be derived from the empirical covariances at lag zero

$$\tilde{g}_0^{\Delta\mathcal{L}} = \tilde{g}_0^{\mathcal{S}} + \tilde{s}_{\mathcal{N}}^2. \tag{5}$$

Thus, it is allowed to split up $\tilde{g}_0^{\Delta\mathcal{L}}$ into the signal variance $\tilde{g}_0^{\mathcal{S}}$ given by the covariance function $g^{\mathcal{S}}(0) = \tilde{g}_0^{\mathcal{S}}$ and a white noise component $\tilde{s}_{\mathcal{N}}^2$, known as the nugget effect (e.g., [9] (p. 59)). In theory, $\tilde{s}_{\mathcal{N}}^2$ can be manually chosen such that the function plausibly decreases from $\tilde{g}_0^{\mathcal{S}}$ towards $\tilde{g}_1^{\mathcal{S}}$ and the higher lags. A more elegant way is to estimate the analytical function $g^S(\tau)$ from empirical covariances $\widetilde{g}_j^S$ with $j>0$ only. Naturally, all estimated functions $g^S(\tau)$ must result in $\widetilde{g}_0^{\Delta \mathcal{L}} - g^S(0) \geq 0$.

Covariance function modeling is a task in various fields of application. They are used for example to represent a stochastic model of the observations within parameter estimation in e.g., laserscanning [18], GPS [47,48], or gravity field modeling [49–54]. The collocation approach is closely related to Gaussian Process Regression from the machine learning domain [55]. The family of covariance functions presented here can be naturally used as kernel functions in such approaches.

Within these kinds of applications, the covariance functions are typically fitted to empirically derived covariances which follow from post-fit residuals. Within an iterative procedure, the stochastic model can be refined. Furthermore, they are used to characterize the signal characteristics, again e.g., in gravity field estimation or geoid determination in the context of least squares collocation [7] or in atmospheric sciences [11,14].

Reference [8] (Ch. 3) proposed to have a special look to only the three parameters, variance, correlation length, and curvature at the origin to fit a covariance function to the empirical covariance sequence. Reference [17] also suggested taking the sequence of zero crossings into account to find an appropriate set of base functions. In addition, most approaches proceed with the fixing of appropriate base functions for the covariance model as a first step. This step corresponds to the determination of the type or family of the covariance function. The fitting then is restricted to this model and does not generalize well to other and more complex cases. Furthermore, it is common to manually fix certain parameters and optimize only a subset parameters in an adjustment procedure; see e.g., [18].

Once the covariance model is fixed, various optimization procedures exist to derive the model parameters which result in a best fit of the model to a set of empirical covariances. Thus, another aspect of covariance function estimation is the numerical implementation of the estimation procedure. Visual strategies versus least squares versus Maximum Likelihood, point cloud versus representative empirical covariance sequences and non robust versus robust estimators are various implementation possibilities discussed in the literature (see e.g., [11,14,18,47,51,56]). To summarize, the general challenges of covariance function fitting are to find an appropriate set of linear independent base functions, i.e., the type of covariance function, and the nonlinear nature of the set of chosen base functions together with the common problem of outlier detection and finding good initial values for the estimation process.

In particular, geodetic data often exhibit negative correlations or even oscillatory behavior in the covariance functions which leaves a rather limited field of types of covariance functions, e.g., cosine and cardinal sine functions in the one-dimensional or Bessel functions in two-dimensional case (e.g., [27]). One general class of covariance functions with oscillatory behavior is discussed in the next section.

3. The Second-Order Gauss–Markov Process

3.1. The Covariance Function of the SOGM-Process

A widely used covariance function is based on the second-order Gauss–Markov processes as given in [23] (Equation (5.2.36)) and [25] (Ch. 4.11, p. 185). The process defines a covariance function of the form

$$\gamma(\tau) = \frac{\sigma^2}{\cos(\eta)} e^{-\zeta \omega_0 \tau} \cos\left(\sqrt{1-\zeta^2}\, \omega_0 \tau - \eta\right)$$
$$= \sigma^2 e^{-\zeta \omega_0 \tau} \left(\cos\left(\sqrt{1-\zeta^2}\, \omega_0 \tau\right) + \tan(\eta) \sin\left(\sqrt{1-\zeta^2}\, \omega_0 \tau\right)\right) \quad (6)$$

with $0 < \zeta < 1$ and $\omega_0 > 0$.

Its shape is defined by three parameters. ω_0 represents a frequency and ζ is related to the attenuation. The phase η can be restricted to the domain $|\eta| < \pi/2$ for logical reasons.

A reparametrization with $c := \zeta \omega_0$ and $a := \sqrt{1-\zeta^2}\, \omega_0$ gives

$$\gamma(\tau) = \frac{\sigma^2}{\cos(\eta)}\, e^{-c\tau} \cos(a\tau - \eta) \qquad (7)$$
$$= \sigma^2 e^{-c\tau} \left(\cos(a\tau) + \tan(\eta) \sin(a\tau)\right) \qquad \text{with} \quad a, c > 0$$

which highlights the shape of a sinusoid. We distinguish between the nominal frequency a and the natural frequency $\omega_0 = a/\sqrt{1-\zeta^2}$. The relative weight of the sine with respect to the cosine term amounts to $w := \tan(\eta)$. It is noted here that the SOGM-process is uniquely defined by three parameters. Here, we will use ω_0, ζ and η as the defining parameters. Of course, the variance σ^2 is a parameter as well. However, as it is just a scale and remains independent of the other three, it is not relevant for the theoretical characteristics of the covariance function.

The described covariance function is referenced by various names, e.g., the second-order shaping filter [57], the general damped oscillation curve [27] (Equation (2.116)), and the underdamped second-order continuous-time bandpass filter [58] (p. 270). In fact, the SOGM represents the most general damped oscillating autocorrelation function built from exponential and trigonometric terms. For example, the function finds application in VLBI analysis [59,60].

3.2. Positive Definiteness of the SOGM-Process

At first glance, it is surprising that a damped sine term is allowed in the definition of the covariance function of the SOGM-process (cf. Equation (6)), as the sine is not positive semi-definite. However, it is shown here that it is in fact a valid covariance function, provided that some conditions on the parameters are fulfilled.

The evaluation concerning the positive semi-definiteness of the second-order Gauss–Markov process can be derived by analyzing the process's Fourier transform as given in [25] (Ch. 4.11, p. 185) and the evaluation of it being non-negative (cf. the Bochner theorem, [61]). The topic is discussed and summarized in [60]. With some natural requirements already enforced by $0 \leq \zeta \leq 1$ and $\omega_0 > 0$, the condition for positive semi-definiteness (cf. e.g., [57] (Equation (A2))) is

$$|\sin(\eta)| \leq \zeta \qquad (8)$$

which can be expressed by the auxiliary variable $\alpha := \arcsin(\zeta)$ as $|\eta| \leq \alpha$.

In terms of the alternative parameters a and c, this condition translates to $w \leq c/a$ (cf. e.g., [27] (Equation (2.117))). As a result, non-positive definite functions as the sine term are allowed in the covariance function only if the relative contribution compared to the corresponding cosine term is small enough.

4. Discrete AR-Processes

4.1. Definition of the Process

A more general and more flexible stochastic process is defined by the autoregressive (AR) process. An AR-process is a time series model which relates signal time series values, or more specifically the signal sequence, S_i with autoregressive coefficients α_k as (e.g., [26] (Ch. 3.5.4))

$$S_i = \sum_{k=1}^{p} \alpha_k S_{i-k} + \mathcal{E}_i\,. \qquad (9)$$

With the transition to $\bar{\alpha}_0 := 1, \bar{\alpha}_1 := -\alpha_1, \bar{\alpha}_2 := -\alpha_2$, etc., the decorrelation relation to white noise $\mathcal{E}_i$ is given by

$$\sum_{k=0}^{p} \bar{\alpha}_k \, \mathcal{S}_{i-k} = \mathcal{E}_i \, . \tag{10}$$

The characteristic polynomial in the factorized form (cf. [26] (Ch. 3.5.4))

$$\bar{\alpha}_0 \, x^p + \bar{\alpha}_1 \, x^{p-1} + \ldots + \bar{\alpha}_p = \prod_{k=1}^{p} (x - p_k) \tag{11}$$

has the roots $p_k \in \mathbb{C}$, i.e., the poles of the autoregressive process, which can be complex numbers. This defines a unique transition between coefficients and poles. In common practice, the poles of an AR(p)-process only appear as a single real pole or as complex conjugate pairs. Following this, an exemplary process of order $p = 4$ can be composed by either two complex conjugate pairs, or one complex pair and two real poles, or four individual single real poles. An odd order gives at least on real pole. For general use, AR-processes are required to be stationary. This requires that its poles are inside the unit circle, i.e., $|p_k| < 1$ (cf. [26] (Ch. 3.5.4)).

AR-processes define an underlying covariance function as well. We will provide it analytically for the AR(2)-process and will summarize a computation strategy for the higher order processes in the following.

4.2. The Covariance Function of the AR(2)-Process

Although AR-processes are defined by discrete covariance sequences, the covariance function can be written as a closed analytic expression, which, evaluated at the discrete lags, gives exactly the discrete covariances. For instance, the AR(2)-process has a covariance function which can be written in analytical form. The variance of AR-processes is mostly parameterized using the variance $\sigma_\mathcal{E}^2$ of the innovation sequence $\mathcal{E}_i$. In this paper, however, we use a representation with the autocorrelation function and a variance σ^2 as in [26] (Equation (3.5.36), p. 130) and [62] (Section 3.5, p. 504). In addition, the autoregressive parameters α_1 and α_2 can be converted to the parameters a and c via $a = \arccos(\alpha_1/(2\sqrt{-\alpha_2}))$ and $c = -\ln(\sqrt{-\alpha_2})$. Hence, the covariance function of the AR(2)-process can be written in a somewhat complicated expression in the variables a and c as

$$\gamma(\tau) = \sigma^2 \sqrt{(\cot(a) \, \tanh(c))^2 + 1} \, e^{-c\tau} \, \cos(a\,\tau - \arctan(\cot(a) \, \tanh(c))) \tag{12}$$

using the phase $\eta = \arctan(\cot(a) \, \tanh(c))$ or likewise

$$\gamma(\tau) = \sigma^2 \, e^{-c\tau} \, (\cos(a\,\tau) + \tanh(c) \, \cot(a) \, \sin(a\,\tau)) \tag{13}$$

with the weight of the sine term $w = \tanh(c) \cot(a)$.

Please note that in contrast to the SOGM-process the weight or phase in Equations (12) and (13) cannot be set independently, but depends on a and c. Thus, this model is defined by two parameters only. Therefore, the SOGM-process is the more general model. Caution must be used with respect to the so-called second-order autoregressive covariance model of [11], which is closely related but does not correspond to the standard discrete AR-process.

4.3. AR(p)-Process

The covariance function of an AR(p)-process is given as a discrete series of covariances $\{\gamma_j\}_{\Delta t}$ defined at discrete lags $h = j\,\Delta t$ with distance Δt. The Yule–Walker equations (e.g., [24] (Section 3.2); [30] (Equation (11.8)))

$$
\begin{array}{c}
(14\text{a}) \\
\\
(14\text{b}) \\
\\
\\
(14\text{c})
\end{array}
\left[
\begin{array}{ccccc}
\gamma_1 & \gamma_2 & \gamma_3 & \cdots & \gamma_p \\
\gamma_0 & \gamma_1 & \gamma_2 & \cdots & \gamma_{p-1} \\
\gamma_1 & \gamma_0 & \gamma_1 & \cdots & \gamma_{p-2} \\
\gamma_2 & \gamma_1 & \gamma_0 & \cdots & \vdots \\
\vdots & \vdots & \vdots & \ddots & \gamma_1 \\
\gamma_{p-1} & \gamma_{p-2} & \cdots & \gamma_1 & \gamma_0 \\
\gamma_p & \gamma_{p-1} & \cdots & \gamma_2 & \gamma_1 \\
\gamma_{p+1} & \gamma_p & \cdots & \gamma_3 & \gamma_2 \\
\vdots & \vdots & \ddots & & \vdots \\
\gamma_{n-1} & \gamma_{n-2} & \cdots & \gamma_{n-p+1} & \gamma_{n-p}
\end{array}
\right]
\left[
\begin{array}{c}
\alpha_1 \\ \alpha_2 \\ \alpha_3 \\ \vdots \\ \alpha_p
\end{array}
\right]
=
\left[
\begin{array}{c}
\gamma_0 - \sigma_\varepsilon^2 \\
\gamma_1 \\
\gamma_2 \\
\gamma_3 \\
\vdots \\
\gamma_p \\
\gamma_{p+1} \\
\gamma_{p+2} \\
\vdots \\
\gamma_n
\end{array}
\right]
\quad (14)
$$

directly relate the covariance sequence to the AR(p) coefficients. With (14a) being the 0th equation, the next p Yule–Walker equations (first to pth, i.e., (14b)) are the linear system mostly used for estimation of the autoregressive coefficients. Note that this system qualifies for the use of Levinson–Durbin algorithm because it is Toeplitz structured, cf. [30] (Ch. 11.3) and [63].

The linear system containing only the higher equations (14c) is called the modified Yule–Walker (MYW) equations, cf. [64]. This defines an overdetermined system which can be used for estimating the AR-process parameters where n lags are included.

The recursive relation

$$\gamma_i - \alpha_1 \gamma_{i-1} - \alpha_2 \gamma_{i-2} - \ldots - \alpha_p \gamma_{i-p} = 0, \quad \text{for} \quad i = p+1, p+2, \ldots \quad (15)$$

represents a pth order linear homogeneous difference equation whose general solution is given by the following equation with respect to a discrete and equidistant time lag $h = |t-t'| = j\Delta t$

$$\gamma_j = A_1 p_1^h + A_2 p_2^h + \ldots + A_p p_p^h \quad \text{with} \quad j \in \mathbb{N}_0,\; h \in \mathbb{R}^+,\; A_k \in \mathbb{C}, \quad (16)$$

cf. [23] (Equation (5.2.44)) and [26] (Equation (3.5.44)). p_k are the poles of the process (cf. Equation (11)) and A_k some unknown coefficients.

It has to be noted here that covariances of AR(p)-processes are generally only defined at discrete lags. However, it can be mathematically shown that the analytic function of Equation (13) exactly corresponds to the covariance function Equation (6) of the SOGM. In other words, the interpolation of the discrete covariances is done using the same sinusoidal functions as in Equation (6) such that the covariance function of the AR(p)-process can equally be written with respect to a continuous time lag $\tau = |t-t'|$ by

$$\gamma(\tau) = \mathrm{Re}\left(A_1 p_1^\tau + A_2 p_2^\tau + \ldots + A_p p_p^\tau\right) = \mathrm{Re}\left(\sum_{k=1}^{p} A_k p_k^\tau\right) \quad \text{with} \quad \tau \in \mathbb{R}^+, A_k \in \mathbb{C}. \quad (17)$$

This is also a valid solution of Equation (15) in the sense that $\gamma(h) = \gamma_j$ holds. For one special case of poles, which are negative real poles, the function can be complex valued due to the continuous argument τ. Thus, the real part has to be taken for general use.

Now, assuming A_k and p_k to be known, Equation (17) can be used to interpolate the covariance defined by an AR-process for any lag τ. Consequently, the covariance definition of an AR-process leads to an analytic covariance function which can be used to interpolate or approximate discrete covariances.

4.3.1. AR(2)-Model

We can investigate the computation for the processes of second order in detail. Exponentiating a complex number mathematically corresponds to

$$p_k^\tau = |p_k|^\tau \left(\cos(\arg(p_k)\,\tau) + i\sin(\arg(p_k)\,\tau) \right). \tag{18}$$

As complex poles always appear in conjugate pairs, it is plausible that for complex conjugate pairs $p_k = p_l^*$ the coefficients A_k and A_l are complex and also conjugate to each other $A_k = A_l^*$. Thus, $A_k p_k^\tau + A_l p_l^\tau$ becomes $A_k p_k^\tau + A_k^*(p_k^*)^\tau$ and the result will be real.

From Equation (13), we can derive that the constants amount to $A_{k,l} = \frac{\sigma^2}{2}\left(1 \pm i \tanh(c)\cot(a)\right) = \frac{\sigma^2}{2}(1 \pm i\,w)$ for the AR(2)-process and from Equation (18) we can see that $c = -\ln(|p_k|)$ and $a = |\arg(p_k)|$ such that the covariance function can be written as

$$\gamma(\tau) = \sigma^2 \sqrt{(\tanh(\ln(|p_k|))\cot(|\arg(p_k)|))^2 + 1} \\ |p_k|^\tau \cos(|\arg(p_k)|\tau + \arctan(\tanh(\ln(|p_k|))\cot(|\arg(p_k)|))) \tag{19}$$

$$= \sigma^2 |p_k|^\tau \left(\cos(|\arg(p_k)|\tau) - \tanh(\ln(|p_k|))\cot(|\arg(p_k)|)\sin(|\arg(p_k)|\tau) \right). \tag{20}$$

It is evident now that the AR(2) covariance model can be expressed as an SOGM covariance function. Whilst the SOGM-process has three independent parameters, here, both damping, frequency, and phase of Equation (19) are determined by only two parameters $|p_k|$ and $|\arg(p_k)|$ based on $e^{-c} = |p_k|$, $c = -\ln(|p_k|)$, $a = |\arg(p_k)|$ and $\eta = \arctan(\tanh(\ln(|p_k|))\cot(|\arg(p_k)|))$. Thus, the SOGM-process is the more general model, whereas the AR(2)-process has a phase η or weight w that is not independent. From Equation (19), phase η can be recovered from the A_k by

$$|\eta_k| = |\arg(A_k)| \tag{21}$$

and the weight by $|w| = |\mathrm{Im}(A_k)/\mathrm{Re}(A_k)|$.

4.3.2. AR(1)-Model

Here, the AR(1)-model appears as a limit case. Exponentiating a positive real pole results in exponentially decaying behavior. Thus, for a single real positive pole, one directly gets the exponential Markov-type AR(1) covariance function, also known in the literature as first-order Gauss–Markov (FOGM), cf. [65] (p. 81). A negative real pole causes discrete covariances of alternating sign. In summary, the AR(1)-process gives the exponentially decaying covariance function for $0 < p_k < 1$

$$\gamma(\tau) = \sigma^2 \exp(-c\,\tau) \quad \text{with} \quad c = -\ln(|p_k|) \\ = \sigma^2 |p_k|^\tau \tag{22}$$

or the exponentially decaying oscillation with Nyquist frequency for $-1 < p_k < 0$, cf. [23] (p. 163), i.e.,

$$\gamma(\tau) = \sigma^2 \exp(-c\,\tau) \cos(\pi\,\tau) \\ = \sigma^2 |p_k|^\tau \cos(\pi\,\tau). \tag{23}$$

4.4. Summary

From Equation (17), one can set up a linear system

$$\begin{bmatrix} \gamma_0 \\ \gamma_1 \\ \gamma_2 \\ \gamma_3 \end{bmatrix} = \begin{bmatrix} p_1^0 & p_2^0 & p_3^0 & p_4^0 \\ p_1^1 & p_2^1 & p_3^1 & p_4^1 \\ p_1^2 & p_2^2 & p_3^2 & p_4^2 \\ p_1^3 & p_2^3 & p_3^3 & p_4^3 \end{bmatrix} \begin{bmatrix} A_1 \\ A_2 \\ A_3 \\ A_4 \end{bmatrix} \quad \text{or} \quad \begin{bmatrix} \gamma_1 \\ \gamma_2 \\ \gamma_3 \\ \gamma_4 \end{bmatrix} = \begin{bmatrix} p_1^1 & p_2^1 & p_3^1 & p_4^1 \\ p_1^2 & p_2^2 & p_3^2 & p_4^2 \\ p_1^3 & p_2^3 & p_3^3 & p_4^3 \\ p_1^4 & p_2^4 & p_3^4 & p_4^4 \end{bmatrix} \begin{bmatrix} A_1 \\ A_2 \\ A_3 \\ A_4 \end{bmatrix}, \quad (24)$$

here shown exemplarily for an AR(4)-process. The solution of Equation (24) uniquely determines the constants A_k applying standard numerical solvers, assuming the poles to be known from the process coefficients, see Equation (11).

Since Equation (17) is a finite sum over exponentiated poles, the covariance function of a general AR(p)-process is a sum of, in case of complex conjugate pairs, AR(2)-processes in the shape of Equation (19) or, in case of real poles, damping terms as given in Equations (22) and (23). The great advantage is that the choice of poles is automatically done by the estimation of the autoregressive process by the YW-Equations (14). Here, we also see that the AR(2)-process as well as both cases of the AR(1)-process can be modeled with Equation (17) such that the proposed approach automatically handles both cases.

Furthermore, we see that Equation (17) adds up the covariance functions of the forms of Equation (19), Equation (22), or Equation (23) for each pole or pair of poles. Any recursive filter can be uniquely dissected into a cascade of second-order recursive filters, described as second-order sections (SOS) or biquadratic filter, cf. [58] (Ch. 11). Correspondingly, the poles of amount p can be grouped into complex-conjugate pairs or single real poles. Thus, the higher order model is achieved by concatenation of the single or paired poles into the set of p poles (vector p) and correspondingly by adding up one SOGM covariance function for each section. Nonetheless, this is automatically done by Equation (17).

5. Generalization to ARMA-Models

5.1. Covariance Representation of ARMA-Processes

Thus far, we introduced fitting procedures for the estimation of autoregressive coefficients as well as a linear system of equations to simply parameterize the covariance function of AR(p)-processes. In this section, we demonstrate that ARMA-models can be handled with the same linear system and the fitting procedure thus generalizes to ARMA-processes.

For the upcoming part, it is crucial to understand that the exponentiation p_k^τ of Equation (17) exactly corresponds to the exponentiation defined in the following way:

$$e^{s_k \tau} = e^{\mathrm{Re}(s_k)\tau} \left(\cos(\mathrm{Im}(s_k)\,\tau) + i \sin(\mathrm{Im}(s_k)\,\tau) \right) \tag{25}$$

i.e., $p_k^\tau = e^{s_k \tau}$, if the transition between the poles p_k to s_k is done by $s_k = \ln(p_k) = (\ln(|p_k|) + i \arg(p_k))$ and $p_k = e^{s_k}$. To be exact, this denotes the transition of the poles from the z-domain to the Laplace-domain. This parametrization of the autoregressive poles can, for example, be found in [23] (Equation (5.2.46)), [66] (Equation (A.2)), and [26] (Equation (3.7.58)). In these references, the covariance function of the AR-process is given as a continuous function with respect to the poles s_k such that the use of Equation (17) as a continuous function is also justified.

In the literature, several parametrizations of the moving average part exist. Here, we analyze the implementation of [33] (Equation (2.15)), where the covariance function of an ARMA-process is given by

$$\gamma(\tau) = \sum_{k=1}^{p} \frac{b(s_k)\,b(-s_k)}{a'(s_k)\,a(-s_k)} e^{s_k \tau}. \tag{26}$$

Inserting $p_k^\tau = e^{s_k \tau}$ and denoting $A_k := \frac{b(s_k)\,b(-s_k)}{a'(s_k)\,a(-s_k)}$ which is independent of τ, we obtain

$$\gamma(\tau) = \sum_{k=1}^{p} \frac{b(s_k)\,b(-s_k)}{a'(s_k)\,a(-s_k)} p_k^\tau = \sum_{k=1}^{p} A_k\, p_k^\tau, \qquad (27)$$

which is suitable for the purpose of understanding the parametrization chosen in this paper. Now, the covariance corresponds to the representation of Equation (17). The equation is constructed by a finite sum of complex exponential functions weighted by a term consisting of some polynomials $a(\,\cdot\,)$, $a'(\,\cdot\,)$ for the AR-part and $b(\,\cdot\,)$ for the MA-part evaluated at the positions of the positive and negative autoregressive roots s_k.

It is evident in Equation (26) that τ is only linked with the poles. This exponentiation of the poles builds the term which is responsible for the damped oscillating or solely damping behavior. The fraction builds the weighting of these oscillations exactly the same way as the A_k in Equation (17). In fact, Equation (17) can undergo a partial fraction decomposition and be represented as in Equation (26). The main conclusion is that ARMA-models can also be realized with Equation (17). The same implication is also gained from the parametrizations by [67] (Equation (3.2)), [31] (Equation (48)), [68] (Equation (9)), and [34] (Equation (4)). It is noted here that the moving average parametrization varies to a great extent in the literature in the sense that very different characteristic equations and zero and pole representations are chosen.

As a result, although the MA-part is extensively more complex than the AR-part and very differently modeled throughout the literature, the MA-parameters solely influence the coefficients A_k weighting the exponential terms, which themselves are solely determined by the autoregressive part. This is congruent with the findings of the Equation (19) where frequency and damping of the SOGM-process are encoded into the autoregressive poles p_k.

5.2. The Numerical Solution for ARMA-Models

Autoregressive processes of order p have the property that $p + 1$ covariances are uniquely given by the process, i.e., by the coefficients and the variance. All higher model-covariances can be recursively computed from the previous ones, cf. Equation (15). This property generally does not hold for empirical covariances, where each covariance typically is an independent estimate. Now, suppose Equation (24) is solved as an overdetermined system by including higher empirical covariances, i.e., covariances that are not recursively defined. The resulting weights A_k will automatically correspond to general ARMA-models because the poles p_k are fixed.

Precisely, the contradiction within the overdetermined system will, to some extent, forcedly end up in the weights A_k and thus in some, for the moment unknown, MA-coefficients. The model still is an SOGM process because the number of poles is still two and the SOGM covariance function is the most general damped oscillating function. The two AR-poles uniquely define the two SOGM-parameters frequency ω_0 and attenuation ζ. The only free parameter to fit an ARMA-model into the shape of the general damped oscillating function (SOGM-process) is the phase η. Hence, the MA-part of arbitrary order will only result in a single weight or phase as in Equation (19) and the whole covariance function can be represented by an SOGM-process. Consequently, the A_k will be different from that of Equation (20), cf. [29] (p. 60), but the phase can still be recovered from Equation (21).

In summary, the general ARMA(2,q)-model (Equation (26)) is also realizable with Equation (17) and thus with the linear system of Equation (24). Here, we repeat the concept of second-order sections. Any ARMA(p,q)-process can be uniquely dissected into ARMA(2,2)-processes. Thus, our parametrization of linear weights to complex exponentials can realize pure AR(2) and general SOGM-processes, which can be denoted as ARMA(2,q)-models. These ARMA(2,q)-processes form single SOGM-processes with corresponding parameters ω_0, ζ and η. The combination of the ARMA(2,q) to the original ARMA(p,q) process is the operation of addition for the covariance (function), concatenation for the poles, and convolution for the coefficients. Thus, the expansion to higher orders is similar to the pure AR(p) case. The finite sum adds up the covariance function for each second-order section which is an SOGM-process.

The weights would have to undergo a partial-fraction decomposition to give the MA-coefficients. Several references exist for decomposing Equation (26) into the MA-coefficients, known as spectral factorization, e.g., by partial fraction decomposition. In this paper, we stay with the simplicity and elegance of Equation (17).

6. Estimation and Interpolation of the Covariance Series

Within this section, the theory summarized above is used for covariance modeling, i.e., estimating covariance functions $g^{\mathcal{S}}(\tau)$, which can be evaluated for any lag τ, from a sequence of given empirical covariances $\{\tilde{g}_j^{\Delta\mathcal{L}}\}_{\Delta t}$. Here, the choice of estimator for the empirical covariances is not discussed and it is left to the user whether to use the biased or the unbiased estimator of the empirical covariances, cf. [23] (p. 174) and [69] (p. 252).

The first step is the estimation of the process coefficients from the $\tilde{g}_j^{\Delta\mathcal{L}}$ with the process order p defined by the user. Furthermore, different linear systems have been discussed for this step, cf. Equation (14), also depending on the choice of n, which is the index of the highest lag included in Equation (14). These choices already have a significant impact on the goodness of fit of the covariance function to the empirical covariances, as will be discussed later. The resulting AR-coefficients α_k can be directly converted to the poles p_k using the factorization of the characteristic polynomial (Equation (11)).

For the second step, based on Equation (16), a linear system of m equations with $m \geq p$, can be set up, but now for the empirical covariances. Using the first m covariances, but ignoring the lag 0 value contaminated by the nugget effect, this results in a system like Equation (24), but now in the empirical covariances $\tilde{g}_j^{\Delta\mathcal{L}} = \tilde{g}_j^{\mathcal{S}}, j > 0$

$$\begin{bmatrix} \tilde{g}_1^{\mathcal{S}} \\ \tilde{g}_2^{\mathcal{S}} \\ \vdots \\ \tilde{g}_{m-1}^{\mathcal{S}} \\ \tilde{g}_m^{\mathcal{S}} \end{bmatrix} = \begin{bmatrix} p_1^1 & p_2^1 & \cdots & p_{p-1}^1 & p_p^1 \\ p_1^2 & p_2^2 & \cdots & p_{p-1}^2 & p_p^2 \\ \vdots & \vdots & & \vdots & \vdots \\ p_1^{m-1} & p_2^{m-1} & \cdots & p_{p-1}^{m-1} & p_p^{m-1} \\ p_1^m & p_2^m & \cdots & p_{p-1}^m & p_p^m \end{bmatrix} \cdot \begin{bmatrix} A_1 \\ A_2 \\ \vdots \\ A_{p-1} \\ A_p \end{bmatrix}. \quad (28)$$

For $m = p$, the system can be uniquely solved, resulting in coefficients A_k which model the covariance function as an AR(p)-process. The case $m > p$ results in a fitting problem of the covariance model to the m empirical covariances $\tilde{g}_j$ with p unknowns. This overdetermined system can be solved for instance in the least squares sense to derive estimates A_k from the $\tilde{g}_j$. As was discussed, these A_k signify a process modeling as an ARMA-model. Here, one could use the notation $\tilde{A}_k$ in order to indicate adjusted parameters in contrast to the uniquely determined A_k for the pure AR-process. For the sake of a unified notation of Equation (17), it is omitted.

Due to the possible nugget effect, it is advised to exclude $\tilde{g}_0^{\Delta\mathcal{L}}$ and solve Equation (28); however, it can also be included, cf. Equation (24). Moreover, a possible procedure can be to generate a plausible $\tilde{g}_0^{\mathcal{S}}$ from a manually determined $\tilde{s}_N^2$ by $\tilde{g}_0^{\mathcal{S}} = \tilde{g}_0^{\Delta\mathcal{L}} - \tilde{s}_N^2$. Equally, the MYW-Equations are a possibility to circumvent using $\tilde{g}_0^{\Delta\mathcal{L}}$.

Modeling Guidelines

The idea of solving the system for the weights A_k is outlined in [23] (p. 167). In the following, we summarize some guidelines to estimate the covariance function starting at the level of some residual observation data.

Initial steps:

- Determine the empirical autocorrelation function $\tilde{g}_0^{\Delta\mathcal{L}}$ to $\tilde{g}_n^{\Delta\mathcal{L}}$ as estimates for the covariances $\gamma_0^{\Delta\mathcal{L}}$ to $\gamma_n^{\Delta\mathcal{L}}$. The biased or unbiased estimate can be used.
- Optional step: Reduce $\tilde{g}_0^{\Delta\mathcal{L}}$ by an arbitrary additive white noise component $\tilde{s}_N^2$ (nugget) such that $\tilde{g}_0^S = \tilde{g}_0^{\Delta\mathcal{L}} - \tilde{s}_N^2$ is a plausible y-intercept to $\tilde{g}_1^S$ and the higher lags.

Estimation of the autoregressive process:

- Define a target order p and compute the autoregressive coefficients α_k by
    - solving the Yule–Walker equations, i.e., Equation (14b), or
    - solving the modified Yule–Walker equations, i.e., Equation (14c), in the least squares sense using $\tilde{g}_1^S$ to $\tilde{g}_n^S$.
- Compute the poles of the process, which follow from the coefficients, see Equation (11). Check if the process is stationary, which requires all $|p_k| < 1$. If this is not given, it can be helpful to make the estimation more overdetermined by increasing n. Otherwise, the target order of the estimation needs to be reduced. A third possibility is to choose only selected process roots and continue the next steps with this subset of poles. An analysis of the process properties such as system frequencies a or ω_0 can be useful, for instance in the pole-zero plot.

Estimation of the weights A_k:

- Define the number of empirical covariances m to be used for the estimation. Set up the linear system cf. Equation (28) either with or without $\tilde{g}_0^S$. Solve the system of equations either
    - uniquely using $m = p$ to determine the A_k. This results in a pure AR(p)-process.
    - or as an overdetermined manner in the least squares sense, i.e., up to $m > p$. This results in an underlying ARMA-process.
- $g^S(0)$ is given by $g^S(0) = \sum_{k=1}^p A_k$ from which $\tilde{s}_N^2$ can be determined by $\tilde{s}_N^2 = \tilde{g}_0^{\Delta\mathcal{L}} - g^S(0)$. If $g^S(0)$ exceeds $\tilde{g}_0^{\Delta\mathcal{L}}$, it is possible to constrain the solution to pass exactly through or below $\tilde{g}_0^{\Delta\mathcal{L}}$. This can be done using a constrained least squares adjustment with the linear condition $\sum_{k=1}^p A_k = \tilde{g}_0^{\Delta\mathcal{L}}$ (cf. e.g., [69] (Ch. 3.2.7)) or by demanding the linear inequality $\sum_{k=1}^p A_k \leq \tilde{g}_0^{\Delta\mathcal{L}}$ [70] (Ch. 3.3–3.5).
- Check for positive definiteness (Equation (8)) of each second-order section (SOGM component). In addition, the phases need to be in the range $|\eta| < \pi/2$. If the solution does not fulfill these requirements, process diagnostics are necessary to determine whether the affected component might be ill-shaped. If the component is entirely negative definite, i.e., with negative $g^S(0)$, it needs to be eliminated.

Here, it also needs to be examined whether the empirical covariances decrease sufficiently towards the high lags. If not, the stationarity of the residuals can be questioned and an enhanced trend reduction might be necessary.

Using the YW-Equations can be advantageous in order to get a unique (well determined) system to be solved for the α_k. By this, one realizes that the analytic covariance function exactly interpolates the first $p+1$ covariances, which supports the fact that they are uniquely given by the process, cf. Equation (14b). On the other hand, including higher lags into the process estimation enables long-range dependencies with lower degree models. The same holds for solving the system of Equation (28) for the A_k using only $m = p$ or $m > p$ covariances. It is left to the user which procedure gives the best fit to the covariances and strongly depends on the characteristics of the data and the application. In summary, there are several possible choices, cf. Table 1.

Table 1. Fitting procedures.

	Equation (28), $m = p$	Equation (28) with $\tilde{g}_0^S$, $m = p - 1$	Equation (28), $m > p$
YW-Equations	AR-model, interpolation of the first $p+1$ covariances	AR-model, approximation	ARMA-model, approximation
MYW-Equations, $n > p$	AR-model, approximation	AR-model, approximation	ARMA-model, approximation

Finally, the evaluation of the analytic covariance function (Equation (17)) can be done by multiplying the same linear system using arbitrary, e.g., dense, τ:

$$\begin{bmatrix} g^S(\tau_1) \\ g^S(\tau_2) \\ \vdots \\ g^S(\tau_n) \end{bmatrix} = \begin{bmatrix} p_1^{\tau_1} & p_2^{\tau_1} & \cdots & p_{p-1}^{\tau_1} & p_p^{\tau_1} \\ p_1^{\tau_2} & p_2^{\tau_2} & \cdots & p_{p-1}^{\tau_2} & p_p^{\tau_2} \\ \vdots & \vdots & & \vdots & \vdots \\ p_1^{\tau_n} & p_2^{\tau_n} & \cdots & p_{p-1}^{\tau_n} & p_p^{\tau_n} \end{bmatrix} \begin{bmatrix} A_1 \\ A_2 \\ \vdots \\ A_{p-1} \\ A_p \end{bmatrix}. \tag{29}$$

Though including complex p_k and A_k, the resulting covariance function values are theoretically real. However, due to limited numeric precision, the complex part might only be numerically zero. Thus, it is advised to eliminate the imaginary part in any case.

7. An Example: Milan Cathedral Deformation Time Series

This example is the well known deformation time series of Milan Cathedral [62]. The time series measurements are levelling heights of a pillar in the period of 1965 to 1977. It is an equidistant time series having 48 values with sampling interval $\Delta t = 0.25$ years. In [62], the time series is deseasonalized and then used for further analysis with autoregressive processes. In this paper, a consistent modeling of the whole signal component without deseasonalization is seen to be advantageous. The time series is detrended using a linear function and the remaining residuals define the stochastic signal, cf. Figure 1.

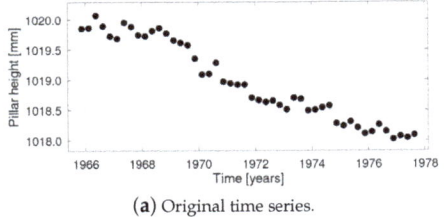
(**a**) Original time series.

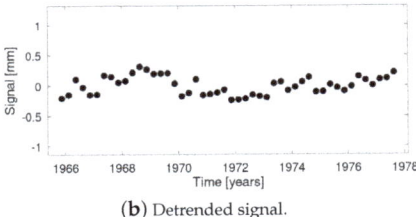
(**b**) Detrended signal.

Figure 1. Milan Cathedral time series.

Based on the detrended time series, the biased estimator is used to determine the empirical covariances in all examples. The order for the AR(p)-process is chosen as $p = 4$. Four different covariance functions were determined based on the method proposed here. All second-order components of the estimated covariance functions are converted to the SOGM parametrization and individually tested for positive semi-definiteness using Equation (8).

As a kind of reference, a manual approach (cf. Figure 2a) was chosen. A single SOGM is adjusted by manually tuning the parameters to achieve a good fit for all empirical covariances, ignoring $\tilde{g}_0^{\Delta \mathcal{L}}$. This function follows the long-wavelength oscillation contained in the signal.

In a second approach, the process coefficients are estimated using the YW-Equations (14b) with the covariances $\tilde{g}_0^{\Delta \mathcal{L}}$ to $\tilde{g}_p^{\Delta \mathcal{L}}$. Furthermore, the system in Equation (28) is solved uniquely using the lags from $\tilde{g}_1^{\Delta \mathcal{L}}$ up to $\tilde{g}_m^{\Delta \mathcal{L}}$ with $m = p$. In Figure 2b, the covariance function exactly interpolates the first five values. This covariance model with relatively low order already contains a second high-frequent oscillation of a 1-year period, caused by annual temperature variations, which is not obvious at first. However, the function misses the long wavelength oscillation and does not fit well to the higher covariances. Here, the model order can of course be increased in order to interpolate more covariances. However, it will be demonstrated now that the covariance structure at hand can in fact also be modeled with the low-order AR(4)-model.

For the remaining examples, the process coefficients are estimated using the MYW-Equations (14c) with $n = 24$ covariances. As a first case, the function was estimated with a manually chosen nugget effect, i.e., $\tilde{g}_0^S = \tilde{g}_0^{\Delta\mathcal{L}} - 0.0019$, and by solving Equation (28) with $m = p - 1$, which results in a pure AR(4)-process. The covariance function represented by Figure 2c approximates the correlations better compared to first two cases. The shape models all oscillations, but does not exactly pass through all values at the lags $\tau = 1.5$ to 4 years.

Finally, the system of Equation (28) was solved using the higher lags $\tilde{g}_m^S$ up to $m = 14$ in order to enforce an ARMA-model. However, the best fit exceeds $\tilde{g}_0^{\Delta\mathcal{L}}$, such that the best valid fit is achieved with a no nugget effect in the signal covariance model and a covariance function exactly interpolating $\tilde{g}_0^{\Delta\mathcal{L}}$. Thus, we fix $g^S(0)$ to $\tilde{g}_0^{\Delta\mathcal{L}}$ using a constrained least squares adjustment, which is the result shown here. In comparison, Figure 2d shows the most flexible covariance function. The function passes very close to nearly all covariances up to $\tau = 6$ years and the fit is still good beyond that. The approximated ARMA-process with order $p = 4$ allows more variation of the covariance function and the function fits better to higher lags.

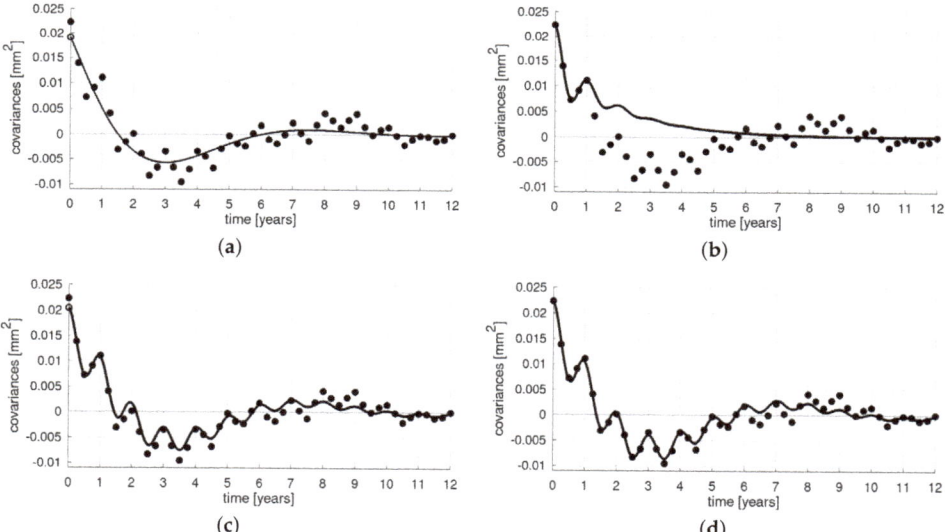

Figure 2. Covariance functions for the Milan Cathedral data sets determined with four different approaches: (**a**) Intuitive ("naive") approach using a single SOGM-process with manually adjusted parameters. (**b**) Covariance function resulting from the interpolation of the first $p + 1$ covariances with a pure AR(4)-processes. (**c**) Covariance function resulting from an approximation procedure of $\tilde{g}_0^S$ to $\tilde{g}_{24}^S$ using a pure AR-process with $p = 4$. (**d**) Covariance function based on approximation with the most flexible ARMA-process ($p = 4$). Empirical covariances are shown by the black dots. The variances $\tilde{g}_0^S$ of the covariance functions are indicated by circles. The parameters of the processes (**c**) and (**d**) are provided in Tables 2 and 3.

The corresponding process parameters for the last two examples are listed in Tables 2 and 3. All parameters are given with numbers in rational approximation. The positive definiteness is directly visible by the condition $|\eta| \leq \alpha$. The approximated pure AR(4)-process in Table 2 is relatively close to being invalid for the second component but still within the limits. For the ARMA-model, notably, poles, frequency, and attenuation parameters are the same, cf. Table 3. Just the phases η of the two components are different and also inside the bounds of positive definiteness. The resulting covariance function using the ARMA-model (Figure 2d) is given by

$$g^S(\tau) = 0.01680 \cdot e^{-0.0723 \frac{\tau}{\Delta t}} \left(\cos\left(0.1950 \frac{\tau}{\Delta t}\right) - 0.3167 \sin\left(0.1950 \frac{\tau}{\Delta t}\right) \right) +$$
$$0.00555 \cdot e^{-0.0653 \frac{\tau}{\Delta t}} \left(\cos\left(1.5662 \frac{\tau}{\Delta t}\right) + 0.02614 \sin\left(1.5662 \frac{\tau}{\Delta t}\right) \right) \quad (30)$$
$$\text{with } \tau \text{ in years.}$$

It needs to be noted here that the connection of unitless poles and parameters with distance τ cannot be done in the way indicated by Sections 2–6. In fact, for the correct covariance function, the argument τ requires a scaling with sampling interval Δt which was omitted for reasons of brevity and comprehensibility. As a consequence, the scaling $1/\Delta t$ to the argument τ is included in Equation (30). We also assess the same factor to the transition from ω_0 to ordinary frequency ν_0 given in Tables 2 and 3.

Table 2. Approximation with pure AR(2)-components. The frequency is also given as ordinary frequency ν_0. The variance of the combined covariance function amounts to $\sigma^2 = 0.02046$ mm² which leads to $\sigma_{\mathcal{N}} = 0.04359$ mm. ω_0 and η are given in units of radians.

	Roots	Frequency ω_0	Frequency ν_0	Damping ζ	Phase η	$\alpha = \arcsin(\zeta)$
A	0.91262 + 0.18022i 0.91262 − 0.18022i	0.20794	0.13238 [1/year]	0.34774	−0.15819	0.35516
B	0.0042636 + 0.93678i 0.0042636 − 0.93678i	1.5676	0.99797 [1/year]	0.041655	0.040221	0.041667

Table 3. Approximation with SOGM-components, i.e., ARMA(2,q)-models. The variance of the combined covariance function amounts to $\sigma^2 = 0.02235$ mm².

	Roots	Frequency ω_0	Frequency ν_0	Damping ζ	Phase η	$\alpha = \arcsin(\zeta)$
A	0.91262 + 0.18022i 0.91262 − 0.18022i	0.20794	0.13238 [1/year]	0.34774	−0.30670	0.35516
B	0.0042636 + 0.93678i 0.0042636 − 0.93678i	1.5676	0.99797 [1/year]	0.041655	0.026133	0.041667

Using the process characteristics in Tables 2 and 3, it is obvious now that the long wavelength oscillation has a period of about 7.6 years. Diagnostics of the estimated process can be done in the same way in order to possibly discard certain components if they are irrelevant. In summary, the proposed approach can realize a much better combined modeling of long and short-wavelength signal components without manual choices of frequencies, amplitudes, and phases. The modified Yule–Walker equations prove valuable for a good fit of the covariance function due to the stabilization by higher lags. ARMA-models provide a further enhanced flexibility of the covariance function.

8. Summary and Conclusions

In this paper, we presented an estimation procedure for covariance functions based on methodology of stochastic processes and a simple and straightforward numerical method. The approach is based on the analogy of the covariance functions defined by the SOGM-process and autoregressive processes. Thus, we provide the most general damped oscillating autocorrelation function built from exponential and trigonometric terms, which includes several simple analytical covariance models as limit cases.

The covariance models of autoregressive process as well as of ARMA-processes correspond to a linear combination of covariance functions of second-order Gauss–Markov (SOGM) processes. We provide fitting procedures of these covariance functions to empirical covariance estimates based on simple systems of linear equations. Notably, the numerical method easily extends to ARMA-processes with the same linear system of equations. In the future, research will be done towards possibilities of constraining the bounds of stationarity and positive definiteness in the estimation steps.

The great advantage is that the method is automated and gives the complete model instead of needing to manually model each component. Our method is very flexible because the process estimation automatically chooses complex or real poles, depending on whether more oscillating or only decaying covariance components are necessary to model the process. Naturally, our approach restricts to stationary time series. In non-stationary cases, the empirically estimated covariance sequence would not decrease with increasing lag and by this contradict the specifications of covariance functions, e.g., $\tilde{g}_0^{\Delta \mathcal{L}}$ being the largest covariance, see Section 2 and [28]. Such an ill-shaped covariance sequence will definitely result in non-stationary autoregressive poles and the method will fail.

The real world example has shown that covariance function estimation can in fact give good fitting results even for complex covariance structures. The guidelines presented here provide multiple possibilities for fitting procedures and process diagnostics. As a result, covariance function estimation is greatly automated with a generic method and a more consistent approach to a complete signal modeling is provided.

Author Contributions: Formal analysis, investigation, validation, visualization, and writing—original draft: T.S., methodology, writing—review and editing: T.S., J.K., J.M.B., and W.-D.S., funding acquisition, project administration, resources, supervision: W.-D.S. All authors have read and agreed to the published version of the manuscript.

Funding: This research was supported by the Deutsche Forschungsgemeinschaft (DFG) Grant No. SCHU 2305/7-1 'Nonstationary stochastic processes in least squares collocation— NonStopLSC'.

Acknowledgments: We thank the anonymous reviewers for their valuable comments.

Conflicts of Interest: The authors declare no conflict of interest.

Abbreviations

The following abbreviations are used in this manuscript:

AR	autoregressive
ARMA	autoregressive moving average
LSC	least squares collocation
MA	moving average
MYW	modified Yule–Walker
SOGM	second-order Gauss–Markov
SOS	second-order sections
YW	Yule–Walker

References

1. Kolmogorov, A.N. *Grundbegriffe der Wahrscheinlichkeitsrechnung*; Springer: Berlin/Heidelberg, Germany, 1933. [CrossRef]
2. Wiener, N. *Extrapolation, Interpolation, and Smoothing of Stationary Time Series*; MIT Press: Cambridge, MA, USA, 1949.
3. Moritz, H. *Advanced Least-Squares Methods*; Number 175 in Reports of the Department of Geodetic Science; Ohio State University Research Foundation: Columbus, OH, USA, 1972.
4. Moritz, H. *Advanced Physical Geodesy*; Wichmann: Karlsruhe, Germany, 1980.
5. Schuh, W.D. Signalverarbeitung in der Physikalischen Geodäsie. In *Handbuch der Geodäsie, Erdmessung und Satellitengeodäsie*; Freeden, W., Rummel, R., Eds.; Springer Reference Naturwissenschaften; Springer: Berlin/Heidelberg, Germany, 2016; pp. 73–121. [CrossRef]

6. Moritz, H. *Least-Squares Collocation*; Number 75 in Reihe A; Deutsche Geodätische Kommission: München, Germany, 1973.
7. Reguzzoni, M.; Sansó, F.; Venuti, G. The Theory of General Kriging, with Applications to the Determination of a Local Geoid. *Geophys. J. Int.* **2005**, *162*, 303–314. [CrossRef]
8. Moritz, H. *Covariance Functions in Least-Squares Collocation*; Number 240 in Reports of the Department of Geodetic Science; Ohio State University: Columbus, OH, USA, 1976.
9. Cressie, N.A.C. *Statistics for Spatial Data*; Wiley Series in Probability and Statistics; Wiley: New York, NY, USA, 1991. [CrossRef]
10. Chilès, J.P.; Delfiner, P. *Geostatistics: Modeling Spatial Uncertainty*; Wiley Series in Probability and Statistics; John Wiley & Sons: Hoboken, NJ, USA, 1999. [CrossRef]
11. Thiébaux, H.J. Anisotropic Correlation Functions for Objective Analysis. *Mon. Weather Rev.* **1976**, *104*, 994–1002. [CrossRef]
12. Franke, R.H. *Covariance Functions for Statistical Interpolation*; Technical Report NPS-53-86-007; Naval Postgraduate School: Monterey, CA, USA, 1986.
13. Gneiting, T. Correlation Functions for Atmospheric Data Analysis. *Q. J. R. Meteorol. Soc.* **1999**, *125*, 2449–2464. [CrossRef]
14. Gneiting, T.; Kleiber, W.; Schlather, M. Matérn Cross-Covariance Functions for Multivariate Random Fields. *J. Am. Stat. Assoc.* **2010**, *105*, 1167–1177. [CrossRef]
15. Meissl, P. *A Study of Covariance Functions Related to the Earth's Disturbing Potential*; Number 151 in Reports of the Department of Geodetic Science; Ohio State University: Columbus, OH, USA, 1971.
16. Tscherning, C.C.; Rapp, R.H. *Closed Covariance Expressions for Gravity Anomalies, Geoid Undulations, and Deflections of the Vertical Implied by Anomaly Degree Variance Models*; Technical Report DGS-208; Ohio State University, Department of Geodetic Science: Columbus, OH, USA, 1974.
17. Mussio, L. Il metodo della collocazione minimi quadrati e le sue applicazioni per l'analisi statistica dei risultati delle compensazioni. In *Ricerche Di Geodesia, Topografia e Fotogrammetria*; CLUP: Milano, Italy, 1984; Volume 4, pp. 305–338.
18. Koch, K.R.; Kuhlmann, H.; Schuh, W.D. Approximating Covariance Matrices Estimated in Multivariate Models by Estimated Auto- and Cross-Covariances. *J. Geod.* **2010**, *84*, 383–397. [CrossRef]
19. Sansò, F.; Schuh, W.D. Finite Covariance Functions. *Bull. Géodésique* **1987**, *61*, 331–347. [CrossRef]
20. Gaspari, G.; Cohn, S.E. Construction of Correlation Functions in Two and Three Dimensions. *Q. J. R. Meteorol. Soc.* **1999**, *125*, 723–757. [CrossRef]
21. Gneiting, T. Compactly Supported Correlation Functions. *J. Multivar. Anal.* **2002**, *83*, 493–508. [CrossRef]
22. Kraiger, G. *Untersuchungen zur Prädiktion nach kleinsten Quadraten mittels empirischer Kovarianzfunktionen unter besonderer Beachtung des Krümmungsparameters*; Number 53 in Mitteilungen der Geodätischen Institute der Technischen Universität Graz; Geodätische Institute der Technischen Universität Graz: Graz, Austria, 1987.
23. Jenkins, G.M.; Watts, D.G. *Spectral Analysis and Its Applications*; Holden-Day: San Francisco, CA, USA, 1968.
24. Box, G.; Jenkins, G. *Time Series Analysis: Forecasting and Control*; Series in Time Series Analysis; Holden-Day: San Francisco, CA, USA, 1970.
25. Maybeck, P.S. *Stochastic Models, Estimation, and Control*; Vol. 141-1, Mathematics in Science and Engineering; Academic Press: New York, NY, USA, 1979. [CrossRef]
26. Priestley, M.B. *Spectral Analysis and Time Series*; Academic Press: London, UK; New York, NY, USA, 1981.
27. Yaglom, A.M. *Correlation Theory of Stationary and Related Random Functions: Volume I: Basic Results*; Springer Series in Statistics; Springer: New York, NY, USA, 1987.
28. Brockwell, P.J.; Davis, R.A. *Time Series Theory and Methods*, 2nd ed.; Springer Series in Statistics; Springer: New York, NY, USA, 1991. [CrossRef]
29. Hamilton, J.D. *Time Series Analysis*; Princeton University Press: Princeton, NJ, USA, 1994.
30. Buttkus, B. *Spectral Analysis and Filter Theory in Applied Geophysics*; Springer: Berlin/Heidelberg, Germany, 2000. [CrossRef]
31. Jones, R.H. Fitting a Continuous Time Autoregression to Discrete Data. In *Applied Time Series Analysis II*; Findley, D.F., Ed.; Academic Press: New York, NY, USA, 1981; pp. 651–682. [CrossRef]
32. Jones, R.H.; Vecchia, A.V. Fitting Continuous ARMA Models to Unequally Spaced Spatial Data. *J. Am. Stat. Assoc.* **1993**, *88*, 947–954. [CrossRef]

33. Brockwell, P.J. Continuous-Time ARMA Processes. In *Stochastic Processes: Theory and Methods*; Shanbhag, D., Rao, C., Eds.; Volume 19, Handbook of Statistics; North-Holland: Amsterdam, The Netherlands, 2001; pp. 249–276. [CrossRef]
34. Kelly, B.C.; Becker, A.C.; Sobolewska, M.; Siemiginowska, A.; Uttley, P. Flexible and Scalable Methods for Quantifying Stochastic Variability in the Era of Massive Time-Domain Astronomical Data Sets. *Astrophys. J.* **2014**, *788*, 33. [CrossRef]
35. Tómasson, H. Some Computational Aspects of Gaussian CARMA Modelling. *Stat. Comput.* **2015**, *25*, 375–387. [CrossRef]
36. Schuh, W.D. *Tailored Numerical Solution Strategies for the Global Determination of the Earth's Gravity Field*; Volume 81, Mitteilungen der Geodätischen Institute; Technische Universität Graz (TUG): Graz, Austria, 1996.
37. Schuh, W.D. The Processing of Band-Limited Measurements; Filtering Techniques in the Least Squares Context and in the Presence of Data Gaps. *Space Sci. Rev.* **2003**, *108*, 67–78. [CrossRef]
38. Klees, R.; Ditmar, P.; Broersen, P. How to Handle Colored Observation Noise in Large Least-Squares Problems. *J. Geod.* **2003**, *76*, 629–640. [CrossRef]
39. Siemes, C. Digital Filtering Algorithms for Decorrelation within Large Least Squares Problems. Ph.D. Thesis, Landwirtschaftliche Fakultät der Universität Bonn, Bonn, Germany, 2008.
40. Krasbutter, I.; Brockmann, J.M.; Kargoll, B.; Schuh, W.D. Adjustment of Digital Filters for Decorrelation of GOCE SGG Data. In *Observation of the System Earth from Space—CHAMP, GRACE, GOCE and Future Missions*; Flechtner, F., Sneeuw, N., Schuh, W.D., Eds.; Vol. 20, Advanced Technologies in Earth Sciences, Geotechnologien Science Report; Springer: Berlin/Heidelberg, Germany, 2014; pp. 109–114. [CrossRef]
41. Farahani, H.H.; Slobbe, D.C.; Klees, R.; Seitz, K. Impact of Accounting for Coloured Noise in Radar Altimetry Data on a Regional Quasi-Geoid Model. *J. Geod.* **2017**, *91*, 97–112. [CrossRef]
42. Schuh, W.D.; Brockmann, J.M. The Numerical Treatment of Covariance Stationary Processes in Least Squares Collocation. In *Handbuch der Geodäsie: 6 Bände*; Freeden, W., Rummel, R., Eds.; Springer Reference Naturwissenschaften; Springer: Berlin/Heidelberg, Germany, 2018; pp. 1–36. [CrossRef]
43. Pail, R.; Bruinsma, S.; Migliaccio, F.; Förste, C.; Goiginger, H.; Schuh, W.D.; Höck, E.; Reguzzoni, M.; Brockmann, J.M.; Abrikosov, O.; et al. First, GOCE Gravity Field Models Derived by Three Different Approaches. *J. Geod.* **2011**, *85*, 819. [CrossRef]
44. Brockmann, J.M.; Zehentner, N.; Höck, E.; Pail, R.; Loth, I.; Mayer-Gürr, T.; Schuh, W.D. EGM_TIM_RL05: An Independent Geoid with Centimeter Accuracy Purely Based on the GOCE Mission. *Geophys. Res. Lett.* **2014**, *41*, 8089–8099. [CrossRef]
45. Schubert, T.; Brockmann, J.M.; Schuh, W.D. Identification of Suspicious Data for Robust Estimation of Stochastic Processes. In *IX Hotine-Marussi Symposium on Mathematical Geodesy*; Sneeuw, N., Novák, P., Crespi, M., Sansò, F., Eds.; International Association of Geodesy Symposia; Springer: Berlin/ Heidelberg, Germany, 2019. [CrossRef]
46. Krarup, T. *A Contribution to the Mathematical Foundation of Physical Geodesy*; Number 44 in Meddelelse; Danish Geodetic Institute: Copenhagen, Denmark, 1969.
47. Amiri-Simkooei, A.; Tiberius, C.; Teunissen, P. Noise Characteristics in High Precision GPS Positioning. In *VI Hotine-Marussi Symposium on Theoretical and Computational Geodesy*; Xu, P., Liu, J., Dermanis, A., Eds.; International Association of Geodesy Symposia; Springer: Berlin/ Heidelberg, Germany, 2008; pp. 280–286. [CrossRef]
48. Kermarrec, G.; Schön, S. On the Matérn Covariance Family: A Proposal for Modeling Temporal Correlations Based on Turbulence Theory. *J. Geod.* **2014**, *88*, 1061–1079. [CrossRef]
49. Tscherning, C.; Knudsen, P.; Forsberg, R. *Description of the GRAVSOFT Package*; Technical Report; Geophysical Institute, University of Copenhagen: Copenhagen, Denmark, 1994.
50. Arabelos, D.; Tscherning, C.C. Globally Covering A-Priori Regional Gravity Covariance Models. *Adv. Geosci.* **2003**, *1*, 143–147. [CrossRef]
51. Arabelos, D.N.; Forsberg, R.; Tscherning, C.C. On the a Priori Estimation of Collocation Error Covariance Functions: A Feasibility Study. *Geophys. J. Int.* **2007**, *170*, 527–533. [CrossRef]
52. Darbeheshti, N.; Featherstone, W.E. Non-Stationary Covariance Function Modelling in 2D Least-Squares Collocation. *J. Geod.* **2009**, *83*, 495–508. [CrossRef]

53. Barzaghi, R.; Borghi, A.; Sona, G. New Covariance Models for Local Applications of Collocation. In *IV Hotine-Marussi Symposium on Mathematical Geodesy*; Benciolini, B., Ed.; International Association of Geodesy Symposia; Springer: Berlin/Heidelberg, Germany, 2001; pp. 91–101. [CrossRef]
54. Kvas, A.; Behzadpour, S.; Ellmer, M.; Klinger, B.; Strasser, S.; Zehentner, N.; Mayer-Gürr, T. ITSG-Grace2018: Overview and Evaluation of a New GRACE-Only Gravity Field Time Series. *J. Geophys. Res. Solid Earth* **2019**, *124*, 9332–9344. [CrossRef]
55. Rasmussen, C.; Williams, C. *Gaussian Processes for Machine Learning*; Adaptive Computation and Machine Learning; MIT Press: Cambridge, MA, USA, 2006.
56. Jarmołowski, W.; Bakuła, M. Precise Estimation of Covariance Parameters in Least-Squares Collocation by Restricted Maximum Likelihood. *Studia Geophysica et Geodaetica* **2014**, *58*, 171–189. [CrossRef]
57. Fitzgerald, R.J. Filtering Horizon-Sensor Measurements for Orbital Navigation. *J. Spacecr. Rocket.* **1967**, *4*, 428–435. [CrossRef]
58. Jackson, L.B. *Digital Filters and Signal Processing*, 3rd ed.; Springer: New York, NY, USA, 1996. [CrossRef]
59. Titov, O.A. Estimation of the Subdiurnal UT1-UTC Variations by the Least Squares Collocation Method. *Astron. Astrophys. Trans.* **2000**, *18*, 779–792. [CrossRef]
60. Halsig, S. Atmospheric Refraction and Turbulence in VLBI Data Analysis. Ph.D. Thesis, Landwirtschaftliche Fakultät der Universität Bonn, Bonn, Germany, 2018.
61. Bochner, S. *Lectures on Fourier Integrals*; Number 42 in Annals of Mathematics Studies; Princeton University Press: Princeton, NJ, USA, 1959.
62. Sansò, F. The Analysis of Time Series with Applications to Geodetic Control Problems. In *Optimization and Design of Geodetic Networks*; Grafarend, E.W., Sansò, F., Eds.; Springer: Berlin/Heidelberg, Germany, 1985; pp. 436–525. [CrossRef]
63. Kay, S.; Marple, S. Spectrum Analysis—A Modern Perspective. *Proc. IEEE* **1981**, *69*, 1380–1419. [CrossRef]
64. Friedlander, B.; Porat, B. The Modified Yule-Walker Method of ARMA Spectral Estimation. *IEEE Trans. Aerosp. Electron. Syst.* **1984**, *AES-20*, 158–173. [CrossRef]
65. Gelb, A. *Applied Optimal Estimation*; The MIT Press: Cambridge, MA, USA, 1974.
66. Phadke, M.S.; Wu, S.M. Modeling of Continuous Stochastic Processes from Discrete Observations with Application to Sunspots Data. *J. Am. Stat. Assoc.* **1974**, *69*, 325–329. [CrossRef]
67. Tunnicliffe Wilson, G. Some Efficient Computational Procedures for High Order ARMA Models. *J. Stat. Comput. Simul.* **1979**, *8*, 301–309. [CrossRef]
68. Woodward, W.A.; Gray, H.L.; Haney, J.R.; Elliott, A.C. Examining Factors to Better Understand Autoregressive Models. *Am. Stat.* **2009**, *63*, 335–342. [CrossRef]
69. Koch, K.R. *Parameter Estimation and Hypothesis Testing in Linear Models*, 2nd ed.; Springer: Berlin/Heidelberg, Germany, 1999. [CrossRef]
70. Roese-Koerner, L. Convex Optimization for Inequality Constrained Adjustment Problems. Ph.D. Thesis, Landwirtschaftliche Fakultät der Universität Bonn, Bonn, Germany, 2015.

 © 2020 by the authors. Licensee MDPI, Basel, Switzerland. This article is an open access article distributed under the terms and conditions of the Creative Commons Attribution (CC BY) license (http://creativecommons.org/licenses/by/4.0/).

Article

Evaluation of VLBI Observations with Sensitivity and Robustness Analyses

Pakize Küreç Nehbit [1,2,*], Robert Heinkelmann [2], Harald Schuh [2,3], Susanne Glaser [2], Susanne Lunz [2], Nicat Mammadaliyev [2,3], Kyriakos Balidakis [2], Haluk Konak [1] and Emine Tanır Kayıkçı [4]

- [1] Department of Geomatics Engineering, Kocaeli University, Kocaeli 41001, Turkey; hkonak@kocaeli.edu.tr
- [2] GFZ German Research Centre for Geosciences, 14473 Potsdam, Germany; robert.heinkelmann@gfz-potsdam.de (R.H.); harald.schuh@tu-berlin.de (H.S.); susanne.glaser@gfz-potsdam.de (S.G.); susanne.lunz@gfz-potsdam.de (S.L.); nicat.mammadaliyev@campus.tu-berlin.de (N.M.); kyriakos.balidakis@gfz-potsdam.de (K.B.)
- [3] Chair of Satellite Geodesy, Technische Universität Berlin, 10623 Berlin, Germany
- [4] Department of Geomatics Engineering, Karadeniz Technical University, Trabzon 61080, Turkey; etanir@ktu.edu.tr
- * Correspondence: nehbit@gfz-potsdam.de

Received: 27 April 2020; Accepted: 3 June 2020; Published: 8 June 2020

Abstract: Very Long Baseline Interferometry (VLBI) plays an indispensable role in the realization of global terrestrial and celestial reference frames and in the determination of the full set of the Earth Orientation Parameters (EOP). The main goal of this research is to assess the quality of the VLBI observations based on the sensitivity and robustness criteria. Sensitivity is defined as the minimum displacement value that can be detected in coordinate unknowns. Robustness describes the deformation strength induced by the maximum undetectable errors with the internal reliability analysis. The location of a VLBI station and the total weights of the observations at the station are most important for the sensitivity analysis. Furthermore, the total observation number of a radio source and the quality of the observations are important for the sensitivity levels of the radio sources. According to the robustness analysis of station coordinates, the worst robustness values are caused by atmospheric delay effects with high temporal and spatial variability. During CONT14, it is determined that FORTLEZA, WESTFORD, and TSUKUB32 have robustness values changing between 0.8 and 1.3 mm, which are significantly worse in comparison to the other stations. The radio sources 0506-612, NRAO150, and 3C345 have worse sensitivity levels compared to other radio sources. It can be concluded that the sensitivity and robustness analysis are reliable measures to obtain high accuracy VLBI solutions.

Keywords: very long baseline interferometry; sensitivity; internal reliability; robustness; CONT14

1. Introduction

Very Long Baseline Interferometry (VLBI) is used to measure the arrival time differences of the signals that come from extragalactic radio sources to antennas separated by up to one Earth diameter. The main principle of the VLBI technique is to observe the same extragalactic radio source synchronously with at least two radio telescopes. Global distances can be measured with millimeter accuracy using the VLBI technique [1,2].

In 1967, for the first time, VLBI was used for the detection of light deflection [3,4]. Nowadays, VLBI is a primary technique to determine global terrestrial reference frames and in particular their scale, celestial reference frame, and the Earth Orientation Parameters (EOP), which consist of universal time, and terrestrial and celestial pole coordinates [2]. Since VLBI is the only technique that connects the celestial with the terrestrial reference frames, the technique is fundamentally different from the other space geodetic techniques. The radio sources are objects in the International Celestial Reference Frame (ICRF); however, the antenna coordinates are obtained in the International Terrestrial Reference Frame (ITRF).

As a result of its important role in either the Celestial Reference Frame or the Terrestrial Reference Frame, it is essential to investigate the quality of VLBI observations and its effect on the unknown parameters. For this reason, the quality of the VLBI observations was investigated according to sensitivity and robustness criteria. Although robustness and sensitivity criteria are not new methods in geodesy, they have been applied to the VLBI observations for the first time in this study. The location of the weak stations and radio sources were easily detected using sensitivity and robustness criteria. Using reliability criteria allows detecting observations that have undetectable gross errors on the unknown parameters. Besides, investigation of the sensitivity of the network against the outliers plays a crucial role in the improvement of accuracy. In this way, the scheduling can be improved using this method in the future.

Sensitivity criteria have been an inspiration for many scientific investigations. The criteria can be explained as the network capability for the monitoring of crustal movements and deformations [5]. So far, mostly geodetic networks based on GPS measurements have been analyzed: sensitivity levels were determined in [6], the datum definition was investigated using sensitivity in [7], a priori sensitivity levels were computed in [8], and the determination of the experimental sensitivity capacities was examined in [9].

Robustness criteria were developed as a geodetic network analysis alternative to standard statistical analysis in [10]. Robustness analysis is the combination of the reliability analysis introduced in [11] and the geometrical strength analysis. Robustness has been the main topic of many studies until today. Different strain models were defined with homogeneous and heterogeneous deformation models in [12]. The displacement vectors defined the effect of the undetectable gross error on the coordinate unknowns, which was determined independently from the translation in [13]. In addition, to obtain the corrected displacement vector, global initial conditions represented by the whole station network were used. Local initial conditions aiming at minimizing the total displacement were developed for the polyhedron represented by each network point as defined in [14].

The paper is organized as follows. Section 2 presents the theoretical background of the sensitivity analysis in the VLBI network. Section 3 introduces the theoretical background of robustness. Section 4 investigates the sensitivity and robustness levels of the VLBI network observed during the continuous campaign CONT14, a continuous VLBI session, which will be further described in Section 4. There, 15 VLBI sessions were evaluated, and the outliers were detected using the software VieVS@GFZ (G2018.7, GFZ, Potsdam, Germany) [15], a fork from the Vienna VLBI Software [16]. The least-squares adjustment module of the VieVS@GFZ software was modified to determine the sensitivity levels of the stations and the radio sources and to obtain the robustness level of the observing stations. The sensitivity levels of the stations and the radio sources were obtained using the developed module for the 15 24-h sessions. The computed sensitivity levels of the stations and radio sources were compared session by session. In addition, the deformation resistance induced by the maximum undetectable errors with the internal reliability analysis was computed for each session. The obtained displacement vectors were compared with threshold values. Conclusions and recommendations for further research will be provided in Section 5.

2. The Sensitivity in the VLBI Network

Sensitivity is the minimum value of undetectable gross errors in the adjusted coordinate differences. The sensitivity levels give information about the weakness of a network. The sensitivity level is computed using the cofactor matrix of the displacement vector estimated from two different sessions.

Using the adjusted coordinates $\hat{x}^1$, $\hat{x}^2$ and their cofactor matrices Q_{xx}^1, Q_{xx}^2 based on different sessions 1 and 2, the displacement vector (d) and the corresponding cofactor matrix (Q_{dd}) for one point (reference point of a station or radio source) are obtained using the following equations:

$$d = \hat{x}^1 - \hat{x}^2 \tag{1}$$

$$Q_{dd} = Q_{xx}^1 + Q_{xx}^2. \tag{2}$$

Alternatively, when it is aimed to obtain the sensitivity level of each session as a priori sensitivity level, the cofactor matrix of the displacement vector is obtained as $Q_{dd} = Q_{xx}$ [9,14] and the weight matrix of the displacement vector for each station P_{d_i} is computed by the following equations

$$\begin{bmatrix} d_1 \\ d_2 \\ \vdots \\ d_n \end{bmatrix} = \begin{bmatrix} N_{11} & N_{12} & \cdot & N_{1n} \\ N_{21} & N_{22} & \cdot & N_{2n} \\ \cdot & \cdot & \cdot & \cdot \\ N_{n1} & N_{n2} & \cdot & N_{nn} \end{bmatrix}^+ \begin{bmatrix} (A^T Pdl)_1 \\ (A^T Pdl)_2 \\ \vdots \\ (A^T Pdl)_n \end{bmatrix} \tag{3}$$

$$d_i = \begin{bmatrix} dx_i \\ dy_i \\ dz_i \end{bmatrix} = \ddot{N}_i A^T Pdl \tag{4}$$

$$Q_{d_i d_i} = \ddot{N}_i A^T P Q_{ll} P A \ddot{N}_i^T = \ddot{N}_i N \ddot{N}_i^T \tag{5}$$

$$P_{d_i} = \left(Q_{d_i d_i} \right)^{-1} \tag{6}$$

where $i = 1, \ldots, n$ is the number of stations, A is the design matrix, P is the weight matrix, N is the normal equation system, $A^T Pdl$ is the right-hand side vector, Q_{ll} is the cofactor matrix of the observations, d_i is the displacement vector at the ith station, P_{d_i} is the weight matrix of the displacement vector at the ith station, and $\ddot{N}_i$ is the sub-matrix of the normal equation system for the ith station.

The obtained weight matrix P_{d_i} is decomposed into its eigenvalue and eigenvector. The minimum detectable displacement value—namely, the best sensitivity level (d_{min})—depends on the inverse of the maximum eigenvalue of the weight matrix (λ_{max}) for each station

$$\| d \|_{min} = \frac{W_0 \sigma}{\sqrt{\lambda_{max}}} \tag{7}$$

where σ is derived from the theoretical variance of the unit weight [6] and the threshold value of the non-centrality parameter (W_0) is determined through $W_0 = W(\alpha_0, \gamma_0, h, \infty)$ based on the power of the test $\gamma_0 = 80\%$, the significance level $\alpha_0 = 5\%$, and the degree of freedom $h = 1$ [17,18].

With single-session VLBI analysis, station and radio source coordinates, clock parameters, pole coordinates, and Universal Time 1 (UT1) minus Universal Time Coordinated (UTC), celestial pole coordinates, and atmosphere parameters can be determined [2,19]. To evaluate the VLBI observations, the mathematical model of the least-squares adjustment is expanded by the matrix of constraints H. The functional model for the actual observations l and constraint parameters l_h can be written as follows:

$$v = Ax - l \tag{8}$$

$$v_c = Hx - l_h \tag{9}$$

where v is the residual vector, v_c is the residual vector of constraints, and x denotes the vector of the unknown parameters [20]. Accordingly, the functional model of the adjustment can be summarized with the following symbolic equations:

$$\begin{bmatrix} v \\ v_c \end{bmatrix} = \begin{bmatrix} A \\ H \end{bmatrix} x - \begin{bmatrix} l \\ l_h \end{bmatrix} \qquad (10)$$

and the corresponding stochastic model of the adjustment is written as:

$$P = \begin{bmatrix} P_{ll} & \\ & P_c \end{bmatrix} \qquad (11)$$

where P_{ll} is the weight matrix of the actual observations, P_c is the weight matrix of the constraint parameters, and the remaining elements of this block-diagonal matrix are equal to zero. According to the adjustment model, the cofactor matrix of the unknown parameters is determined as:

$$Q_{xx} = \left(A^T P A + H^T P_c H \right)^{-1} \qquad (12)$$

where Q_{xx} covers all unknown parameters of the respective VLBI session.

Using the functional and the stochastic models, the unknown parameters are computed with a free network adjustment. The cofactor matrix of the displacement vector of each station is as follows

$$Q_{xx} = \begin{bmatrix} \cdot & \cdot & \cdot & \cdot & \cdot & \cdot & \cdot & \cdot \\ \cdot & q_{x1x1} & q_{x1x2} & q_{x1y1} & q_{x1y2} & q_{x1z1} & q_{x1z2} & \cdot \\ \cdot & q_{x2x1} & q_{x2x2} & q_{x2y1} & q_{x2y2} & q_{x2z1} & q_{x2z2} & \cdot \\ \cdot & q_{y1x1} & q_{y1x2} & q_{y1y1} & q_{y1y2} & q_{y1z1} & q_{y1z2} & \cdot \\ \cdot & q_{y2x1} & q_{y2x2} & q_{y2y1} & q_{y2y2} & q_{y2z1} & q_{y2z2} & \cdot \\ \cdot & q_{z1x1} & q_{z1x2} & q_{z1y1} & q_{z1y2} & q_{z1z1} & q_{z1z2} & \cdot \\ \cdot & q_{z2x1} & q_{z2x2} & q_{z2y1} & q_{z2y2} & q_{z2z1} & q_{z2z2} & \cdot \\ \cdot & \cdot & \cdot & \cdot & \cdot & \cdot & \cdot & \cdot \end{bmatrix}_{u,u} \qquad (13)$$

where u is the number of unknown parameters. For the first station, the matrix $\ddot{N}_i$ is determined as

$$\ddot{N}_1 = \begin{bmatrix} \cdot\cdot & \cdot\cdot & \cdot\cdot & q_{x1x1} & q_{x1x2} & q_{x1y1} & q_{x1y2} & q_{x1z1} & q_{x1z2} \\ \cdot\cdot & \cdot\cdot & \cdot\cdot & q_{y1x1} & q_{y1x2} & q_{y1y1} & q_{y1y2} & q_{y1z1} & q_{y1z2} \\ \cdot\cdot & \cdot\cdot & \cdot\cdot & q_{z1x1} & q_{z1x2} & q_{z1y1} & q_{z1y2} & q_{z1z1} & q_{z1z2} \end{bmatrix}_{3,u} \qquad (14)$$

and the cofactor matrix of the displacement vector of the first station is obtained as

$$\left(Q_{d_1 d_1} \right)_{3,3} = \ddot{N}_1 N \ddot{N}_1^T = \begin{bmatrix} q_{x1x1} & q_{x1y1} & q_{x1z1} \\ q_{y1x1} & q_{y1y1} & q_{y1z1} \\ q_{z1x1} & q_{z1y1} & q_{z1z1} \end{bmatrix}_{3,3}. \qquad (15)$$

In analogy, for each radio source, the cofactor matrix of the displacement vector is obtained using the following equations

$$Q_{xx} = \begin{bmatrix} \cdot & \cdot & \cdot & \cdot & \cdot & \cdot \\ \cdot & q_{\alpha 1 \alpha 1} & q_{\alpha 1 \alpha 2} & q_{\alpha 1 \delta 1} & q_{\alpha 1 \delta 2} & \cdot \\ \cdot & q_{\alpha 2 \alpha 1} & q_{\alpha 2 \alpha 2} & q_{\alpha 2 \delta 1} & q_{\alpha 2 \delta 2} & \cdot \\ \cdot & q_{\delta 1 \alpha 1} & q_{\delta 1 \alpha 2} & q_{\delta 1 \delta 1} & q_{\delta 1 \delta 2} & \cdot \\ \cdot & q_{\delta 2 \alpha 1} & q_{\delta 2 \alpha 2} & q_{\delta 2 \delta 1} & q_{\delta 2 \delta 2} & \cdot \\ \cdot & \cdot & \cdot & \cdot & \cdot & \cdot \end{bmatrix}_{u,u} \qquad (16)$$

α and δ are the source equatorial coordinates defined as right ascension and declination, respectively, and u is defined as above. For the first radio source, the matrix $\ddot{N}_i$ is determined as

$$\ddot{N}_1 = \begin{bmatrix} \cdot & \cdot & q_{\alpha 1 \alpha 1} & q_{\alpha 1 \alpha 2} & q_{\alpha 1 \delta 1} & q_{\alpha 1 \delta 2} & \cdot & \cdot \\ \cdot & \cdot & q_{\delta 1 \alpha 1} & q_{\delta 1 \alpha 2} & q_{\delta 1 \delta 1} & q_{\delta 1 \delta 2} & \cdot & \cdot \end{bmatrix}_{2,u} \tag{17}$$

and the cofactor matrix of the displacement vector of the first radio source is

$$\left(Q_{d_1 d_1}\right)_{2,2} = \ddot{N}_1 N \ddot{N}_1^T = \begin{bmatrix} q_{\alpha 1 \alpha 1} & q_{\alpha 1 \delta 1} \\ q_{\delta 1 \alpha 1} & q_{\delta 1 \delta 1} \end{bmatrix}_{2,2}. \tag{18}$$

Subsequently, the corresponding weight matrix belonging to each station or radio source is computed as shown in Equation (6), and the minimum value of the undetectable gross errors is found by Equation (7).

3. The Robustness of VLBI Stations

Robustness is defined as a function of the reliability criteria [10]. On the other hand, the robustness of a geodetic network is defined as the strength of deformation caused by undetectable gross errors with the internal reliability analysis. The robustness analysis consists of enhancing the internal reliability analysis with the strain technique [10,21].

Internal reliability can be interpreted as the controlling of an observation via the other observations in a network. It can be quantified as the magnitude of the undetectable gross errors by using hypothesis testing. For correlated observations, the internal reliability of the jth observation is obtained with the following equations:

$$\Delta_{0j} = m_0 \sqrt{\frac{W_0}{e_j^T P Q_{\hat{v}\hat{v}} P e_j}} \tag{19}$$

$$e_j^T = \begin{bmatrix} .. & 0 & 0 & 1 & 0 & ... \end{bmatrix} \tag{20}$$

where m_0 is derived from the a posteriori value of the experimental variance, $Q_{\hat{v}\hat{v}}$ is the cofactor matrix of the residuals, e_j^T is a selection vector, which consists of 1 for the jth observation and 0 for the other observations; its dimension equals the total number of observations.

The robustness of each VLBI station is quantified as the effect of the maximal undetectable gross error on the coordinate unknowns (Δx) [10,13,22] as

$$\Delta x = Q A^T P \Delta_{0j} \tag{21}$$

$$\Delta_{0j}^T = \begin{bmatrix} .. & 0 & 0 & \delta_{0j} & 0 & ... \end{bmatrix} \tag{22}$$

where Δ_{0j}^T is a vector, which consists of the internal reliability value of the jth observation and 0 for the other observations, with the dimensions of the total number of observations. The displacement vector can be written as

$$\Delta x_i = \begin{bmatrix} \Delta x_i \\ \Delta y_i \\ \Delta z_i \end{bmatrix} = \begin{bmatrix} u_i \\ v_i \\ w_i \end{bmatrix} \tag{23}$$

where u_i, v_i, and w_i are the displacement vector components in the x-, y-, and z-directions.

$$\Delta x_i^T = \begin{bmatrix} \Delta x_1; & \Delta x_2; & \ldots\ldots\ldots\ldots; & \Delta x_j \end{bmatrix} \tag{24}$$

The effect of the undetected gross error on the unknown coordinate is calculated for any coordinate unknown. The effect can be obtained many times using each observation for any coordinate unknown.

Each observation causes strain with different magnitude and direction. For this reason, the observation having maximum effect on the coordinate unknowns must be identified

$$\Delta x_{0j} = max\{|\Delta x_j|\}. \tag{25}$$

It is supposed that the observation having a maximum vector norm causes maximum strain. To compute the vector norm of each observation, the L1 norm is used as

$$\|\Delta x_j\| = |\Delta x_1| + |\Delta x_2| + \ldots + |\Delta x_u| \tag{26}$$

where u is the number of unknowns.

For the strain computation, the surface formed by the station and its neighboring stations, which are connected through baselines, is used. The strain resulting from the effect of the undetectable gross errors on the coordinate unknowns can be obtained for the polyhedron represented by each network point, with affine or extended Helmert transformation models [14,22].

The displacement vector related to the strain parameters can be determined with the equations:

$$\begin{bmatrix} \Delta x_i \\ \Delta y_i \\ \Delta z_i \end{bmatrix} = E_i \begin{bmatrix} X_i - X_0 \\ Y_i - Y_0 \\ Z_i - Z_0 \end{bmatrix} \tag{27}$$

$$E_i = \begin{bmatrix} \frac{\partial u}{\partial x} & \frac{\partial u}{\partial y} & \frac{\partial u}{\partial z} \\ \frac{\partial v}{\partial x} & \frac{\partial v}{\partial y} & \frac{\partial v}{\partial z} \\ \frac{\partial w}{\partial x} & \frac{\partial w}{\partial y} & \frac{\partial w}{\partial z} \end{bmatrix} = \begin{bmatrix} e_{xx} & e_{xy} & e_{xz} \\ e_{yx} & e_{yy} & e_{yz} \\ e_{zx} & e_{zy} & e_{zz} \end{bmatrix} \tag{28}$$

where E_i is the strain tensor, X_0, Y_0, and Z_0 are the initial conditions, X_i, Y_i, and Z_i are the coordinate unknowns of the ith station located on the surface, e_{xx} is the rate of change in the x-direction with respect to the position component in the x-direction [12].

The strain parameters are independent of the location of surfaces in the coordinate system. For this reason, at each surface, the strain tensor is computed with a reference point P_0 selected on the surface. Using the obtained strain tensor, the objective function is linearized according to the initial conditions via

$$\sum_{i=1}^{n} (\Delta x)^T E_i^T E_i (\Delta x) \rightarrow min \tag{29}$$

$$\sum_{i=1}^{n} E_i^T E_i (X_i - X_0) = 0 \tag{30}$$

$$-\sum_{i=1}^{n} E_i^T E_i X_0 + \sum_{i=1}^{n} E_i^T E_i X_i = 0 \tag{31}$$

where the initial conditions $X_0^T = [X_0 \quad Y_0 \quad Z_0]$ are computed as follows

$$X_0 = \left[\sum_{i=1}^{n} E_i^T E_i \right]^{-1} \sum_{i=1}^{n} E_i^T E_i X_i . \tag{32}$$

When inserting these initial conditions into Equation (27), the corrected displacement vector is obtained [13]. In other words, the displacement vector is translated to the gravity center of the surface computed as:

$$d_i = \sqrt{u_i^2 + v_i^2 + w_i^2}. \tag{33}$$

If the corrected displacement vector is estimated from the surface represented by the whole network of stations, the corrected global displacement vector is obtained. However, if the corrected displacement vector is estimated from the surface represented by each station, local initial conditions (X_{L0}) are obtained as

$$X_{L0} = \left(E_i^T E_i\right)^{-1} E_i^T E_i \sum_{i=1}^{m} X_i \tag{34}$$

where m is the number of stations that have a baseline to the ith station. Using the local initial conditions, the corrected local displacement vector is computed via Equation (27). The computed magnitudes of the displacement vectors are compared with the threshold value estimated from confidence ellipsoids [23]:

$$W_i = m_0 \sqrt{3 F_{h,f,1-\alpha_0} trace(Q_{xx})} \tag{35}$$

where f is the degree of freedom, and α_0 is the significance level.

In case of $d_i > W_i$, we conclude that the network station is not robust [13]. In other words, the network is not sensitive enough to possible outliers and their disturbing effects on the coordinate unknowns.

Due to the fact that the displacement vectors obtained for any station represent the effect of undetectable errors on the coordinate unknowns [14,22], the displacement vectors can be compared to the sensitivity levels d_{min} as well.

4. Results

In order to compare VLBI stations and radio sources approximately under the same conditions, such as scheduling and station geographical distribution, we selected the continuous VLBI Campaign 2014 (CONT14) (https://ivscc.gsfc.nasa.gov/program/cont14) for the numerical test. CONT14 consists of 15 continuous VLBI sessions observed from 2014-May-6, 00:00 UT to 2014-May-20, 23:59 UT. The observations of CONT14 were evaluated session by session with the software VieVS@GFZ written in MatLab©.

To obtain the sensitivity levels of the radio sources and stations, Equations (6), (7), and (13)–(18) mentioned in Section 2 and, to obtain the robustness values of the network stations, Equations (19)–(26) mentioned in Section 3 were added to the least-squares adjustment module in VieVS@GFZ.

In order to obtain the strain parameters on the surfaces, displacement vector components and observed baselines were computed with a small C++ program for each session. According to the strain parameters, magnitudes of the corrected local and global displacement vectors were determined for each station and compared with the threshold values.

4.1. Results of the Sensitivity Analysis

The sensitivity level of a station reflects the total observation weights of the station and the remoteness of the station in the network. A small sensitivity value indicates that a station is strongly controlled by the other stations and hence, its sensitivity level is better.

According to the sensitivity analysis of the CONT14 campaign, the subset of European stations, ONSALA60, WETTZELL, ZELENCHK, MATERA, YEBES40M, and partly NYALES20 have the best sensitivity levels based on all sessions, whereas BADARY provides the worst sensitivity level based on all sessions (Figure 1). Across the sessions, there are small but significant differences as well.

The sensitivity levels of the radio sources show that some radio sources in individual sessions have orders of magnitude larger sensitivity levels, e.g., NRAO150, 3C345, 3C454.3, and 0506-612 (Figure 2).

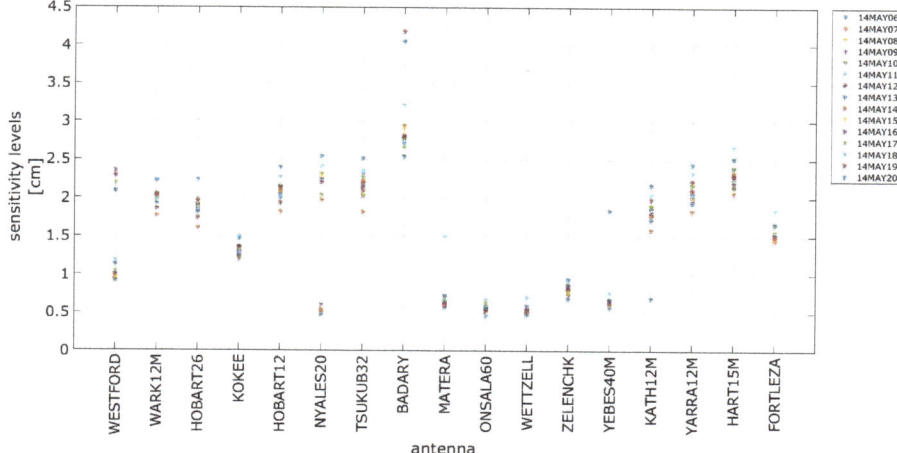

Figure 1. The sensitivity level distributions of the antennas in continuous VLBI Campaign 2014 (CONT14). On the horizontal axis, the antennas are displayed in their respective order of appearance, i.e., unsorted.

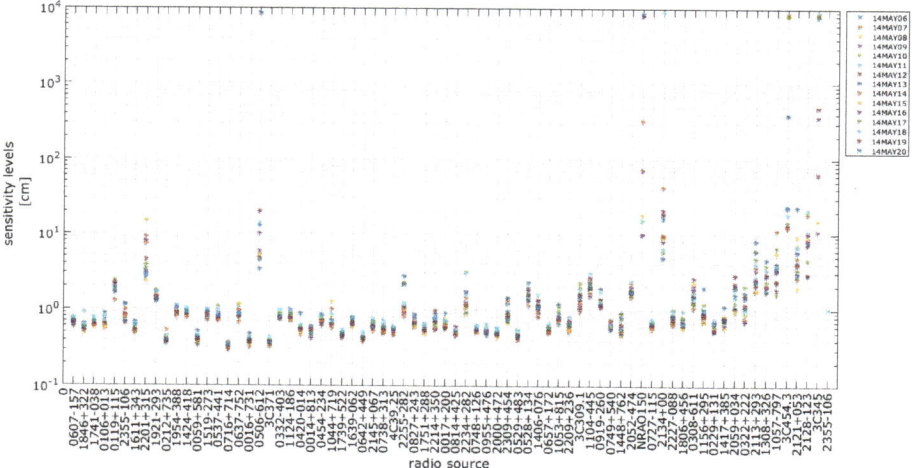

Figure 2. The sensitivity distributions of the radio sources in CONT14. Sources appear randomly on the horizontal axis.

4.2. Result of the Robustness Analysis

The robustness of the network is computed based on the internal reliability and reflects the maximum effect of the undetectable gross error on the coordinate unknowns. In well-designed geodetic networks, the internal reliability value of the observations can be expected below $8m_i$, which is defined as the average error of the observations [8,24–27].

In each session, all observations were tested regarding whether they have gross errors. After the outliers were detected and removed from the observation list, the internal reliability of the observations was investigated.

In Figure 3, some internal reliability values with very large magnitudes can be easily identified. To investigate the large internal reliability magnitudes, the radio sources (and baselines) involved in the observations were identified (Table 1). Comparing the findings to the sensitivity of the radio sources, it could be seen that these radio sources also had the worst sensitivity magnitudes.

If an acceptable mathematical model is used for the adjustment, the statistical analyses can be obtained confidently. For this reason, internal reliability and sensitivity analysis should be performed for all observations.

After all observations of the radio sources mentioned in Table 1 were excluded, it was found that the remaining internal reliabilities fell into a significantly smaller range in Figure 4 compared to Figure 3. Using the outlier-free radio source list, the sensitivity level of the radio sources was obtained. It is seen that radio source 3C454.3 has the maximum sensitivity level (Figure 5). In order to investigate the robustness of the stations with best quality observations, the radio sources having the worst sensitivity levels were excluded. When the observations are reduced according to both internal reliability and the sensitivity levels, the internal reliability criteria can be obtained for the well-designed network.

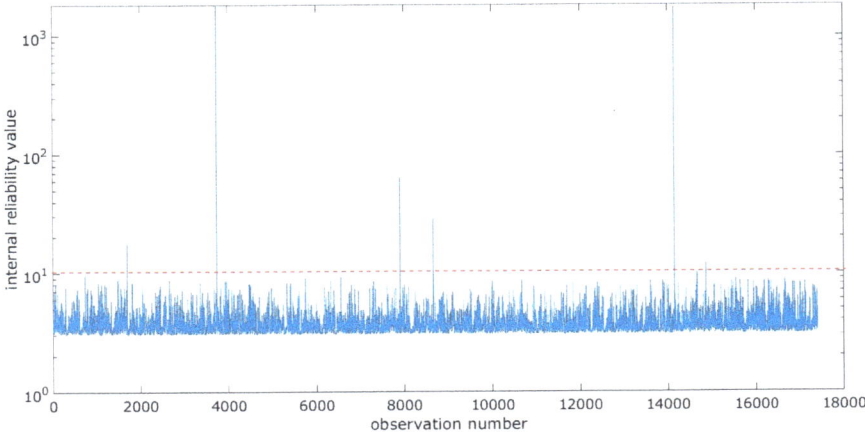

Figure 3. Internal reliability values of the observations in session 14MAY08XA (CONT14) (observations exceeding the red line were identified as outliers and excluded).

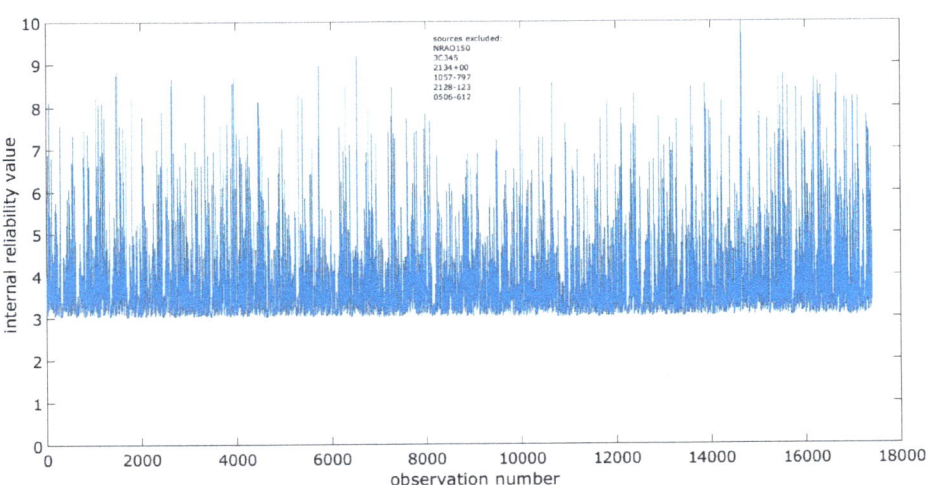

Figure 4. Internal reliability values of the outlier-free observations in session 14MAY08XA.

Table 1. Observations with largest internal reliability values in session 14MAY08XA.

Session	Observation Number	Internal Reliability	Baseline	Radio Source
14MAY08XA	3735	1.83×10^3	MATERA-YEBES40M	NRAO150 (0355+508)
	14,144	1.71×10^3	MATERA-ZELENCHK	3C345 (1641+399)
	7903	6.35×10^1	ONSALA60-ZELENCHK	2134+00 (2134+004)
	1702	1.75×10^1	FORTLEZA-HART15M	1057−797
	8661	2.83×10^1	WESTFORD-FORTLEZA	2128−123
	14,855	1.18×10^1	KATH12M-YARRA12M	0506−612

After this step, the robustness values of the stations were computed. For this purpose, the observation having maximum effect on the coordinate unknowns in each session was selected for the robustness analysis. According to Table 2, it is clearly seen in all sessions that the FORTLEZA station is affected. Table 2 also displays the radio sources that were involved in the observations affecting FORTLEZA. However, the radio sources are identified as rather compact sources because of their small CARMS (closure amplitude rms) values based on natural weighting [28], which are below 0.4.

As mentioned above, the maximum effect of undetected gross error on the station coordinates is called a displacement vector, and it was computed using Equation (21) for CONT14. According to the obtained displacement vector components for CONT14, the magnitudes of the displacement vector components in both x and y directions are about the same but with a different sign, whereas the magnitude in the z direction is about one order of magnitude smaller. In all sessions, FORTLEZA is the most affected one due to undetected gross errors. If we focus on the motion of FORTLEZA during CONT14, the x component of the displacement vector was found to be about between 2 and 4 mm (Figure 6).

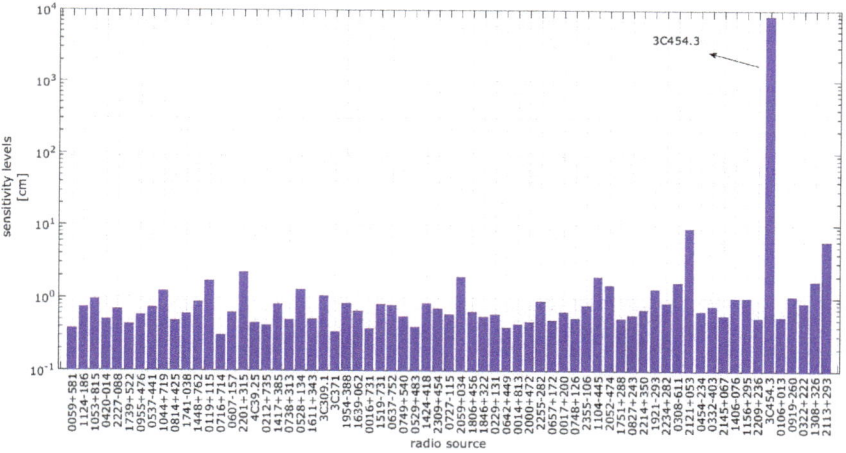

Figure 5. The sensitivity distribution of the radio sources after outlier elimination in session 14MAY08XA.

In each session, the robustness of the stations was obtained with the displacement vector components. To obtain the strain parameters, the surface that was used consists of the station and its neighboring stations connected through baselines. The strain parameters were computed using Equation (27) for the surface that contains each antenna. Using the strain parameters computed for all surfaces represented by the stations, the local displacement vectors were translated according to the gravity center of the surfaces with Equations (27) and (34). The distributions of the local displacement vector magnitudes are illustrated in Figure 7 for one session only: 14MAY08XA.

Table 2. List of observations with a maximum effect of the undetectable gross error on the station coordinates distribution during CONT14.

Session	Observation Number	Baseline	Baseline Length (km)	Affected Station	Radio Source	CARMS Nat. Weight. [28]
14MAY06XA	8973	FORTLEZA-ZELENCHK	8649	FORTLEZA	0454−234	0.17
14MAY07XA	2370	FORTLEZA-HART15M	7025	FORTLEZA	1057−797	0.20
14MAY08XA	13,887	FORTLEZA-WESTFORD	5897	FORTLEZA	0119+115	0.24
14MAY09XA	2207	FORTLEZA-HART15M	7025	FORTLEZA	1057−797	0.20
14MAY10XA	14,313	FORTLEZA-HART15M	7025	FORTLEZA	1424−418	0.18
14MAY11XA	12,560	FORTLEZA-HART15M	7025	FORTLEZA	1424−418	0.18
14MAY12XA	9264	FORTLEZA-HART15M	7025	FORTLEZA	0727−115	0.14
14MAY13XA	15,509	FORTLEZA-WESTFORD	5897	FORTLEZA	0420−014	0.21
14MAY14XA	6772	FORTLEZA-TSUKUB32	12252	FORTLEZA	1611+343	0.36
14MAY15XA	16,477	FORTLEZA-HART15M	7025	FORTLEZA	1751+288	0.18
14MAY16XA	8241	FORTLEZA-HART15M	7025	FORTLEZA	0454−234	0.17
14MAY17XA	5831	FORTLEZA-HART15M	7025	FORTLEZA	0308−611	0.40
14MAY18XA	14,080	FORTLEZA-KATH12M	12553	FORTLEZA	1424−418	0.18
14MAY19XA	1746	FORTLEZA-HART15M	7025	FORTLEZA	1057−797	0.20
14MAY20XA	35	FORTLEZA-WETSFORD	5897	FORTLEZA	0727−115	0.14

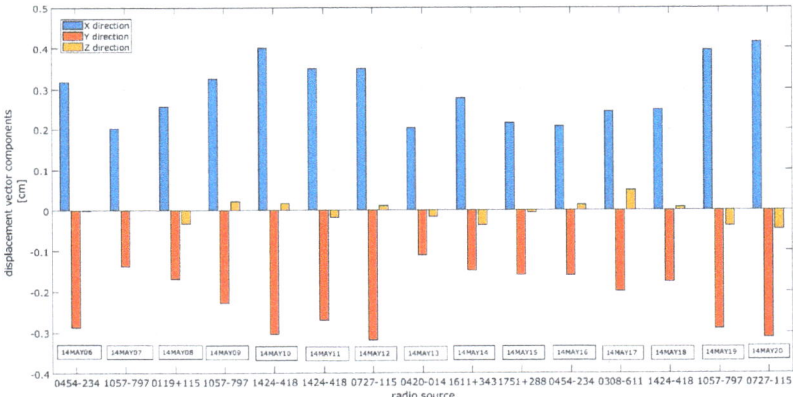

Figure 6. The effect of the undetectable gross errors on station coordinates of FORTLEZA.

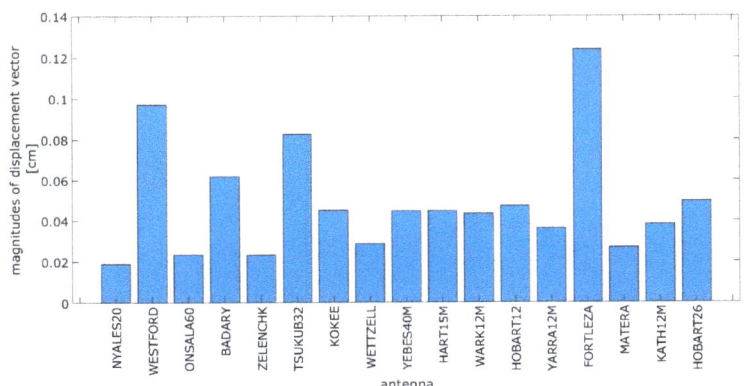

Figure 7. Distribution of local displacement vector magnitudes for Very Long Baseline Interferometry (VLBI) antennas in session 14MAY08XA.

According to Figure 7, FORTLEZA, WESTFORD, and TSUKUB32 have the largest displacement vector magnitudes ranging between 0.8 and 1.3 mm. It can be easily seen that these antennas are affected by the observation having the maximum effect of the undetectable gross errors on the station coordinates.

To address the robustness of the antennas, the computed local displacement vector values were compared to the threshold values obtained applying Equation (35) and the sensitivity levels of the stations as obtained with Equation (7). It was found that all the stations are robust against undetectable gross errors, since the magnitudes of the local displacement vectors are smaller than the threshold values (Figure 8).

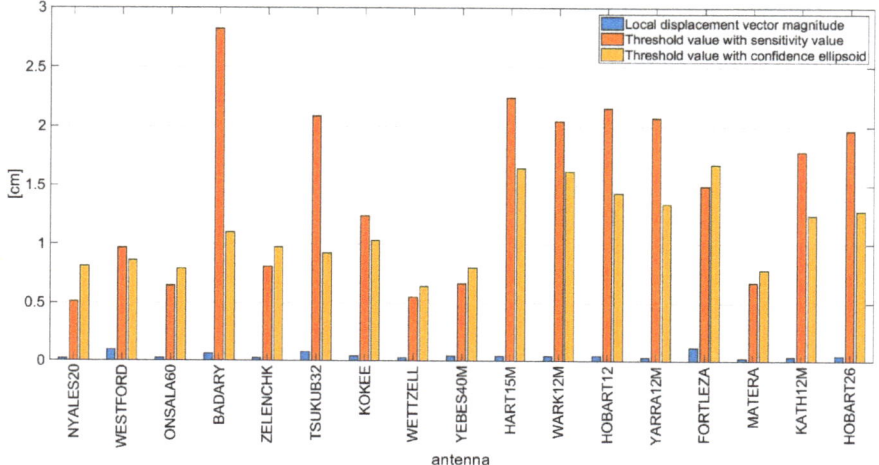

Figure 8. Robustness analysis for VLBI stations in session 14MAY08XA.

5. Discussion

In an astronomical aspect, the structure of the radio sources can cause errors [29], and the astrometric quality of the radio sources is defined by the structure index. Sources with an X-band index of 1 or 2 and S-band index of 1 can be considered as sources of the best astrometric quality. Furthermore, it is recommended that sources with an X-band index of 3 or 4 should not be used [30]. Besides that, previous studies indicate that source structure is a major error source in geodetic VLBI [31]. Sources such as 3C345, 2128−123, and 2134+004 having observations with larger internal reliability values compared to the other sources have a structure index (http://bvid.astrophy.u-bordeaux.fr/) of 4.13, 4.56, and 3.73, respectively, in the X-band. In addition, radio source NRAO150 with a structure index of 2.06 in the S-band has also observations with larger internal reliability. If the radio sources are compared in the view of their sensitivity levels and structure indices, it can be easily understood that the radio source 3C454.3 having a larger sensitivity level has a structure index of 2.9 in the S-band and of 3.84 in the X-band. The radio source 3C454.3 is defined as a quasi-stellar object (QSO) with a core-jet structure that elongates toward the west and bends toward the north-west.

In the robustness analysis, an observation having a maximum effect on the coordinate unknowns more seriously affects those stations used for observing it and their neighboring stations connected with baselines than the other stations in the network. For this reason, network geometry and observation plans are substantial for the robustness analysis. According to the result of the robustness analysis, the observation on the FORTLEZA–WESTFORD baseline has a maximum effect on the coordinate unknowns. In other words, larger magnitudes of the displacement vectors at these stations are obtained. As a result, both FORTLEZA and WESTFORD stations have larger robustness values. In addition, because of the network geometry and the observation plan, TSUKUB32 has larger robustness value than the other stations.

Although the selection of CONT14 is convenient for an initial analysis, the measurements may have systematic errors that cannot be detected in the error analysis because of the short duration of the campaign. Therefore, sensitivity levels of the antennas and the radio sources and robustness values of VLBI antennas may be determined too optimistically.

According to our results, the internal reliability values of the observations and the sensitivity levels of the sources can be used to investigate the source quality together with the structure index. The sources can be excluded based on their sensitivity levels and structure indices. For this reason, it can be considered that robustness and sensitivity criteria can play a substantial role in scheduling in the future.

The software VieVS@GFZ was modified to determine the sensitivity levels and to detect the observations that are having a maximum effect on the coordinate unknowns. It can be used easily for routine analysis of VLBI sessions. However, to obtain the strain parameters and the robustness analysis, VieVS@GFZ should be further modified in the future.

6. Conclusions

In this research, we performed a quality assessment of VLBI observations during CONT14. The radio sources and the VLBI stations that took part in the CONT14 sessions were analyzed according to their sensitivity levels. Furthermore, a robustness analysis was applied for the antennas.

The controllability of one station through the other stations can be investigated by the sensitivity analysis. The location of the station in the network and the total weights of its observations are the most important contributors for the sensitivity. On the other hand, the total observation number of a radio source, and the quality of the observations are also important for the sensitivity levels of the radio sources. It was also found that the investigation of the relationship between the structure of radio source and their sensitivity level is of interest.

According to the robustness analysis of the station coordinates, all of the stations are robust against undetectable gross errors. Some of the stations such as FORTLEZA, WESTFORD, and TSUKUB32 have significantly worse robustness in comparison to the other stations. It is possible that the worst robustness values can be due to the effects of the atmosphere that changes very much with time and with the location of the stations. Another explanation could be the remoteness of the station in the network.

Author Contributions: Analyzing and writing—original draft preparation, P.K.N.; review and editing, R.H., H.S., S.G., S.L., N.M., K.B., H.K., E.T.K. All authors have read and agreed to the published version of the manuscript.

Funding: This research received no external funding.

Acknowledgments: P.K.N. acknowledges the Scientific and Technological Research Council of Turkey (TÜBİTAK) for the financial support of the post-doctoral research program (2219). The International VLBI Service for Geodesy and Astrometry (IVS), [2] and [32], are acknowledged for providing data used within this study. We would like to thank all reviewers for the detailed comments which helped to improve the manuscript.

Conflicts of Interest: The authors declare no conflict of interest.

References

1. GFZ VLBI Group. Available online: https://www.gfz-potsdam.de/en/section/space-geodetic-techniques/topics/geodetic-and-astrometric-vlbi/ (accessed on 2 March 2020).
2. Schuh, H.; Behrend, D. VLBI: A fascinating technique for geodesy and astronomy. *J. Geodyn.* **2012**, *61*, 68–80. [CrossRef]
3. Heinkelmann, R.; Schuh, H. Very long baseline interferometry: Accuracy limits and relativistic tests. *Proc. Int. Astron. Union* **2009**, *5*, 286–290. [CrossRef]
4. Shapiro, I.I. New Method for the Detection of Light Deflection by Solar Gravity. *Science* **1967**, *157*, 806–808. [CrossRef] [PubMed]

5. Even-Tzur, G. Sensitivity Design for Monitoring Deformation Networks. *Boll. Geod. Sci. Affin.* **1999**, *54*, 313–324.
6. Hsu, R.; Hsiao, K. Pre-Computing the Sensitivity of a GPS station for crustal deformation monitoring. *J. Chin. Inst. Eng.* **2002**, *25*, 715–722. [CrossRef]
7. Even-Tzur, G. Datum Definition and Its Influence on the Sensitivity of Geodetic Monitoring Networks. In Proceedings of the 12th FIG Symposium, Baden, Austria, 22–24 May 2006.
8. Küreç, P.; Konak, H. A priori sensitivity analysis for densification GPS networks and their capacities of crustal deformation monitoring: A real GPS network application. *Nat. Hazards Earth Syst. Sci.* **2014**, *14*, 1299–1308. [CrossRef]
9. Konak, H.; Küreç Nehbit, P.; İnce, C.D. Determination of experimental sensitivity capacities against crustal movements of dense-geodetic networks: A real GPS network application. *Arab. J. Geosci.* **2019**, *12*, 596. [CrossRef]
10. Vanicek, P.; Krakiwsky, E.J.; Craymer, M.R.; Gao, Y.; Ong, P.S. *Robustness Analysis, Final Contract Report*; Department of Surveying Engineering Technical Report No. 156; University of New Brunswick: Fredericton, NB, Canada, 1990.
11. Baarda, W. Reliability and precision of networks. In Proceedings of the VII International Course for Engineering Surveys of High Precision, Darmstadt, Germany, 29 September–8 October 1976.
12. Kuang, S. Optimization and Design of Deformation Monitoring Schemes. Ph.D. Thesis, Department of Surveying Engineering Technical Report. University of New Brunswick, Fredericton, NB, Canada, 1991.
13. Berber, M. Robustness Analysis of Geodetic Networks. Ph.D. Thesis, University of Brunswick, Fredericton, NB, Canada, 2006.
14. Küreç Nehbit, P. A Continuous Strain Monitoring and Quality Assessment Strategy for GPS/GNSS Networks. Ph.D. Thesis, University of Kocaeli, Kocaeli, Turkey, 2018.
15. Nilsson, T.; Soja, B.; Karbon, M.; Heinkelmann, R.; Schuh, H. Application of Kalman filtering in VLBI data analysis. *Earth Planets Space* **2015**, *67*, 1–9. [CrossRef]
16. Böhm, J.; Böhm, S.; Nilsson, T.; Pany, A.; Plank, L.; Spicakova, H.; Teke, K.; Schuh, H. The New Vienna VLBI Software VieVS. In *Geodesy for Planet Earth*; Kenyon, S., Pacino, M., Marti, U., Eds.; International Association of Geodesy Symposia; Springer: Berlin/Heidelberg, Germany, 2012; Volume 136. [CrossRef]
17. Bardaa, W. *A Testing Procedure for Use in Geodetic Networks*; Netherlands Geodetic Commission, Publications on Geodesy, Computing Centre of the Delft Geodetic Institute: Delft, The Netherlands, 1968; Volume 2.
18. Aydin, C.; Demirel, H. Computation of Baarda's lower bound of the non-centrality parameter. *J. Geod.* **2004**, *78*, 437–441. [CrossRef]
19. Nothnagel, A. Very Long Baseline Interferometry. In *Handbuch der Geodäsie, 1–58*; Freeden, W., Rummel, R., Eds.; Springer Reference Naturwissenschaften Book Series; Springer Spektrum: Berlin/Heidelberg, Germany, 2018. [CrossRef]
20. Teke, K. Sub-Daily Parameter Estimation in VLBI Data Analysis. Ph.D. Thesis, Institute of Geodesy and Geophysics, Vienna University of Technology, Vienna, Austria, 2011; p. 149.
21. Vanicek, P.; Craymer, M.R.; Krakiwsky, E.J. Robustness analysis of geodetic horizontal networks. *J. Geod.* **2001**, *75*, 199–209.
22. Küreç Nehbit, P.; Konak, H. The global and local robustness analysis in geodetic networks. *Int. J. Eng. Geosci.* **2020**, *5*, 42–48. [CrossRef]
23. Wolf, P.R.; Ghilani, C.D. *Adjustment Computations: Statistics and Least Squares in Surveying and GIS*; John Wiley and Sons, Inc.: Hoboken, NJ, USA, 1997.
24. Baarda, W. Measures for the accuracy of geodetic networks. In Proceedings of the Symposium on Optimization of Design and Computation of Control Networks, Sopron, Hungary, 4–10 July 1977; pp. 419–436.
25. Öztürk, E.; Serbetçi, M. *Adjustment Volume 3*; Technical University of Karadeniz: Trabzon, Turkey, 1992.
26. Yalçınkaya, M.; Teke, K. Optimization of GPS networks with respect to accuracy and reliability criteria. In Proceedings of the 12th FIG Symposium, Baden, Austria, 22–24 May 2006.
27. Pelzer, H. *Geodaetische Netze in Landes- und Ingenieurvermessung II: Vortraege des Kontaktstudiums Februar 1985 in Hannover*; Wittwer: Stuttgart, Germany, 1985.
28. Xu, M.H.; Anderson, J.M.; Heinkelmann, R.; Lunz, S.; Schuh, H.; Wang, G.L. Structure effects for 3417 Celestial Reference Frame radio sources. *Astrophys. J. Suppl. Ser.* **2019**, *242*, 5. [CrossRef]

29. Charlot, P. Radio source structure in Astrometric and geodetic very long baseline interferometry. *Astron. J.* **1990**, *99*, 1309. [CrossRef]
30. Fey, A.L.; Charlot, P. VLBA observations of radio reference frame sources. II. Astrometric suitability based on observed structure. *Astrophys. J. Suppl. Ser.* **1997**, *111*, 95–142. [CrossRef]
31. Anderson, J.M.; Xu, M.H. Source structure and measurement noise are as important as all other residual sources in geodetic VLBI combined. *J. Geophys. Res. Solid Earth* **2018**, *123*, 10–162. [CrossRef]
32. Nothnagel, A.; Artz, T.; Behrend, D.; Malkin, Z. International VLBI service for Geodesy and Astrometry— Delivering high-quality products and embarking on observations of the next generation. *J. Geod* **2017**, *91*, 711–721. [CrossRef]

© 2020 by the authors. Licensee MDPI, Basel, Switzerland. This article is an open access article distributed under the terms and conditions of the Creative Commons Attribution (CC BY) license (http://creativecommons.org/licenses/by/4.0/).

Article

Variance Reduction of Sequential Monte Carlo Approach for GNSS Phase Bias Estimation

Yumiao Tian [1,*], Maorong Ge [2,3,*] and Frank Neitzel [2,*]

1. Faculty of Geosciences and Environmental Engineering, Southwest Jiaotong University, Chengdu 611756, China
2. Institute of Geodesy and Geoinformation Science, Technische Universität Berlin, 10623 Berlin, Germany
3. Section 1.1: Space Geodetic Techniques, GFZ German Research Centre for Geosciences, 14473 Potsdam, Germany
* Correspondence: tymr@163.com (Y.T.); maorong.ge@gfz-potsdam.de (M.G.); frank.neitzel@tu-berlin.de (F.N.)

Received: 29 February 2020; Accepted: 18 March 2020; Published: 3 April 2020

Abstract: Global navigation satellite systems (GNSS) are an important tool for positioning, navigation, and timing (PNT) services. The fast and high-precision GNSS data processing relies on reliable integer ambiguity fixing, whose performance depends on phase bias estimation. However, the mathematic model of GNSS phase bias estimation encounters the rank-deficiency problem, making bias estimation a difficult task. Combining the Monte-Carlo-based methods and GNSS data processing procedure can overcome the problem and provide fast-converging bias estimates. The variance reduction of the estimation algorithm has the potential to improve the accuracy of the estimates and is meaningful for precise and efficient PNT services. In this paper, firstly, we present the difficulty in phase bias estimation and introduce the sequential quasi-Monte Carlo (SQMC) method, then develop the SQMC-based GNSS phase bias estimation algorithm, and investigate the effects of the low-discrepancy sequence on variance reduction. Experiments with practical data show that the low-discrepancy sequence in the algorithm can significantly reduce the standard deviation of the estimates and shorten the convergence time of the filtering.

Keywords: GNSS phase bias; sequential quasi-Monte Carlo; variance reduction

1. Introduction

Global navigation satellite systems (GNSS) are widely used in positioning, navigation, and timing (PNT) services. The accuracy of the precise positioning can reach the level of centimeters and satisfy a pervasive use in civil and military applications. GNSS is being developed at a fast pace, and the systems in full operation at present include the United States of America (USA)'s Global Positioning System (GPS) and Russia's Global Navigation Satellite System (GLONASS). The European Union plans to complete the construction of the Galileo system, while China is going to fully operate the BeiDou Navigation Satellite System (BDS) by 2020 [1,2]. The basic principle of GNSS data processing is to mathematically solve the interesting PNT parameters in the observation models with measurements of the distances between GNSS satellites and receivers. However, the biases in the signal measurements lead to errors in the models and degrade the accuracy of the solutions. Consequently, the bias estimation plays an important role in the quality of the final PNT services [3–5]. Reducing the variance of the bias estimates can more precisely recover the measurements and improve the service quality.

Fast and precise GNSS data processing uses the carrier-wave phase measurement by the receivers. The phase measurement only records the fractional part of the carrier phase plus the cumulated numbers. Therefore, the phase measurements from GNSS receivers are not directly the satellite–receiver distance, and an additional unknown integer ambiguity needs to be solved so that the distance can be recovered. Methods for solving the integer ambiguities were investigated in the past few decades, and some effective approaches such as the ambiguity function method (AFM) and the Least-squares ambiguity Decorrelation Adjustment (LAMBDA) method were proposed, which are widely used in practice [6,7]. The LAMBDA method-based GNSS data processing is usually composed of four steps [6,8]. Firstly, the interesting parameters for PNT services are estimated together with the unknown ambiguities by the least-squares method or Kalman filtering. Secondly, the integer ambiguities are resolved according to the float estimates and variance–covariance (VC) matrix by decorrelation and searching methods. Thirdly, the searched integer ambiguities are validated to assure that they are the correct integers. Fourthly, the interesting unknown parameters of PNT services in the measurement models are derived with the validated integer ambiguities. The reliable ambiguity resolution is critical for fast and precise PNT services. The above steps work well when the errors of the measurements are small, but the performance degrades quickly when the errors grow. The errors of the phase measurements affect the solutions of the float ambiguities in the first step and destroy the integer nature of the ambiguities that are searched in the second step. As a result, the fixed integer vector cannot pass the validation in the third step and, thus, the fast and precise GNSS PNT services will be unreachable.

GNSS signals propagate through the device hardware when they are emitted from the satellite or received by the receivers, leading to time delays, i.e., biases. The biases play the role of errors in the measurements when they cannot be successfully estimated and, thus, block the success of ambiguity fixing. The difficulty in the phase bias estimation lies in the correlation between the unknown bias and the ambiguity parameters. This correlation leads to rank-deficiency of the equation set and, thus, the parameters cannot be solved by the least-squares method or Kalman filtering method in the first step of ambiguity fixing [9,10]. It should be noted that estimation of some parameters leading to rank-deficiency in GNSS data processing can be avoid by the techniques such as S-system theory [11]. However, those techniques focus on solving the estimable parameters and cannot solve the problems when the inestimable parameters are critical. In this case, if we want to estimate the bias parameter or the ambiguity parameter accurately, the conventionally inestimable parameter must be accurately known, which is a dilemma for GNSS researchers. Fortunately, the Monte Carlo-based approaches have the potential to solve this dilemma [12,13]. Furthermore, it can be found in references that the Monte Carlo method is also used for the ambiguity resolution without phase error estimation in attitude determination [14] and code multipath mitigation with only code observations [15]. Those researches use different ideas in data processing and are not related to the topic of phase bias estimation.

The sequential Monte Carlo (SMC) method or particle filtering is used in the state-space approach for time series modeling since the basic procedure proposed by Goden [16] (see the descriptions in References [16–18]). The SMC is mainly to solve problems with non-Gaussian and non-linear models, while it is rarely used in GNSS data processing. SMC can be regarded as Bayesian filtering implemented by the Monte Carlo method [19]. A state-space model, i.e., hidden Markov model, can be described by two stochastic processes $\{x_t\}_{t=1}^T$ and $\{y_t\}_{t=1}^T$. The latent Markov process of initial density satisfies $x_0 \sim \mu(x_0)$, and the Markov transition density is $f(x_k|x_{k-1})$, with $\{y\}_{t=1}^T$ satisfy $g(y_k|x_k)$, which is a conditional marginal density. Bayesian filtering gives the estimation of the posterior density $P(x_k|y_{1:k}) = g(y_k|x_k)g(x_k|x_{k-1})/P(y_k|y_{1:k-1})$, where $P(y_k|y_{1:k-1})$ is a normalizing constant. The analytical solution $P(x_k|y_{1:k})$ can be derived for some special cases such as the solution of a Kalman filter for linear models with Gaussian noise. Otherwise, the analytical solution is not available, and the Monte Carlo-based solutions can be used to approximate the solution via random samples as SMC. The probability density of the variable is represented by weighted samples, and the estimates can be expressed by $\bar{x}_k = 1/N \sum_{i=1}^N x_k^i$. The SMC is mainly composed of three steps according to References [20,21], the update step which updates the weighs of the particles according to $g(y_k|x_k)$, the resampling step to avoid degeneracy indicating most particles with weights close to zero, and the prediction step which transits the particles to the next epoch.

However, the random sequence used in the Monte Carlo method has possible gaps and clusters. Quasi-Monte Carlo (QMC) replaces the random sequence with a low-discrepancy sequence which can reduce the variance and has better performance [22–25]. Until now, the QMC-based variance reduction method in GNSS data processing was not addressed. This study aims to combine the GNSS data processing procedure and the sequential QMC (SQMC) methods to obtain precise GNSS phase bias estimates. The paper firstly gives an overview of the mathematical problem in GNSS bias estimation and then provides a renewed algorithm introducing the variance reduction method based on QMC to precisely estimate GNSS phase bias.

The remainder of this article is structured as follows: Section 2 presents the procedure and mathematical models of the GNSS data processing and introduce the difficulties in phase bias estimation. Section 3 gives an overview of the QMC theory. Section 4 describes the proposed GNSS phase bias estimation algorithm based on the SQMC method. Section 5 gives the results of phase bias estimation with practical data, and Section 6 draws the key research conclusions.

2. Mathematic Models of GNSS Precise Data Processing and the Dilemma

The code and phase measurements are usually used to derive the interesting parameters for GNSS services. The code measurement is the range obtained by multiplying the traveling time of the signal when it propagates from the satellite to the receiver at the speed of light. The phase measurement is much more precise than the code measurement but is ambiguous by an unknown integer number of wavelengths when used as range, and the ambiguities are different every time the receiver relocks the satellite signals [7].

In the measurement models, the unknowns include not only the coordinate parameters, but also the time delays caused by the atmosphere and device hardware, as well as the ambiguities for phase measurement. In relative positioning, the hardware delays can be nonzero values and should be considered in multi-GNSS and GLONASS data processing, i.e., inter-system bias (ISB) [9,10,26] and inter-frequency bias (IFB) [27], respectively. The ISB and IFB of the measurements are correlated with the ambiguities and are the key problems to be solved.

The double difference (DD) measurement models are usually constructed to mitigate the common errors of two non-difference measurements. For short baselines, the DD measurement mathematical models including the interesting parameters for GNSS PNT services, such as coordinates for positioning, the unknown ambiguities, and the ISB or IFB parameters, can be written in the form of matrices as

$$v = Ax + Db + Cy + l, \tag{1}$$

where v denotes the vector of observation residuals; b is composed of unknown single difference (SD) ambiguities $\left(N_{ab}^{i1}, N_{ab}^{i2}, \ldots, N_{ab}^{in}\right)$, where i is the reference satellite and n is the number of the DD-equations, and a and b are the stations; y includes the ISB and IFB rate; vector x contains the unknown station coordinate and the other interesting parameters; l is the measurements from the receiver; A is the design matrix of the elements in x; D is the design matrix with elements of zero and the corresponding carrier wavelength. Matrix D transforms SD ambiguities to DD ambiguities; C is the design matrix of y with elements of zero and the SD of the channel numbers for phase IFB rate parameter, with elements of zero and one for the phase ISB parameter.

GNSS data processing such as for precise positioning is used to precisely determine the elements in x. Denoting the weight matrix of the DD measurements [7] by P, the normal equation of the least-squares method is

$$\begin{bmatrix} A^TPA & A^TPD & A^TPC \\ & D^TPD & D^TPC \\ sym & & C^TPC \end{bmatrix} \begin{bmatrix} x \\ b \\ y \end{bmatrix} = \begin{bmatrix} A^TPl \\ D^TPl \\ C^TPl \end{bmatrix}. \tag{2}$$

For simplification, the notation in Equation (3) is used.

$$\begin{bmatrix} N_{xx} & N_{xb} & N_{xy} \\ & N_{bb} & N_{by} \\ sym & & N_{yy} \end{bmatrix} \begin{bmatrix} x \\ b \\ y \end{bmatrix} = \begin{bmatrix} W_x \\ W_b \\ W_y \end{bmatrix}. \tag{3}$$

If the bias vector y is precisely known, the estimation of x can be realized by following four steps.

Step 1: Derive the solution of x and b with float SD-ambiguities by least-squares method.

$$\begin{bmatrix} \hat{x} \\ \hat{b} \end{bmatrix} = \begin{bmatrix} N_{xx} & N_{xb} \\ N_{bx} & N_{bb} \end{bmatrix}^{-1} \begin{bmatrix} W_x - N_{xy}y \\ W_b - N_{by}y \end{bmatrix} = \begin{bmatrix} Q_{xx} & Q_{xb} \\ Q_{bx} & Q_{bb} \end{bmatrix} \begin{bmatrix} W_x - N_{xy}y \\ W_b - N_{by}y \end{bmatrix}. \tag{4}$$

After the float SD ambiguities in b are estimated, the SD ambiguities and their VC matrix are transformed into DD ambiguities $b_{\hat{D}\hat{D}}$ and the corresponding VC matrix $Q_{\hat{b}\hat{b}}$ by differencing.

Step 2: Fix the integer ambiguities. The elements of $\hat{b}_{DD}$ are intrinsically integer values but the values calculated are floats. Resolving the float values to integers can improve the accuracy to sub-centimeter level with fewer observations [28]. The ambiguity resolution can be expressed by

$$\check{b} = F(\hat{b}), \tag{5}$$

where function $F()$ maps the ambiguities from float values to integers. This process can be implemented by the LAMBDA method [6,8] which can efficiently mechanize the integer least square (ILS) procedure [29]. This method is to solve the ILS problem described by

$$\min(\hat{b} - \overline{b})^T Q_{\hat{b}\hat{b}}^{-1}(\hat{b} - \overline{b}), \text{ with } \overline{b} \in Z^n, \tag{6}$$

where $\overline{b}$ denotes the vector of integer ambiguity candidates. The LAMBDA procedure contains mainly two steps, the reduction step and the search step. The reduction step decorrelates the elements in $\hat{b}$ and orders the diagonal entries by Z-transformations to shrink the search space. The search step is a searching process finding the optimal ambiguity candidates in a hyper-ellipsoidal space.

Step 3: Validate the integer ambiguities. The obtained ambiguity combination $\check{b}$ is not guaranteed to be the correct integer ambiguity vector and it requires to be validated. The R-ratio test [29,30] can be employed. This test tries to ensure that the best ambiguity combination, which is the optimal solution of Equation (6), is statistically better than the second best one. The ratio value is calculated by

$$RATIO = \|\hat{b} - \check{b}'\|^2 Q_{\hat{b}\hat{b}} / \left(\|\hat{b} - \check{b}\|^2 Q_{\hat{b}\hat{b}}\right), \tag{7}$$

where $\check{b}'$ is the second best ambiguity vector according to Equation (6). The integer ambiguity vector $\check{b}$ will be accepted if the ratio value is equal to or larger than a threshold, and it will be refused if the ratio value is smaller than the threshold.

Step 4: Derive the fixed baseline solution. After the integer ambiguity vector passes the validation test, $\check{b}$ is used to adjust the float solution of other parameters, leading to the corresponding fixed solution. This process can be expressed by

$$\check{x} = \hat{x} - Q_{\hat{x}\hat{b}} Q_{\hat{b}\hat{b}}^{-1}(\hat{b} - \check{b}), \tag{8}$$

$$Q_{\check{x}\check{x}} = Q_{\hat{x}\hat{x}} - Q_{\hat{x}\hat{b}} Q_{\hat{b}\hat{b}}^{-1} Q_{\hat{b}\hat{x}}, \tag{9}$$

where $\check{x}$ denotes the fixed solution of x; $Q_{\check{x}\check{x}}$ is the VC matrix of the fixed solution $\check{x}$; $Q_{\hat{b}\hat{x}}$ is the VC matrix of $\hat{b}$ and $\hat{x}$; $\hat{b}$ refers to the float ambiguity solution; $\hat{x}$ is the float solution of x.

The fixed solution $\hat{x}$ can reach sub-centimeter level. If errors in the observation models are removed, the successful ambiguity fixing requires observations of only a few epochs, even a single epoch.

From Equations (5) and (6), the successful ambiguity resolution requires accurate float ambiguity estimates and the corresponding VC matrix. If the bias in y is unknown, the bias cannot be separated by Equation (4) but stays with the ambiguity parameter. As a result, the obtained ambiguity parameters include bias and the ambiguity resolution will fail. When both the bias and the ambiguity are parameterized simultaneously, the bias parameter is correlated with the ambiguity parameter and, thus, it is impossible to separate the bias values and get precise float ambiguities estimates. Mathematically, the normal equation set (Equation (2)) will be rank-deficient and cannot be solved.

3. Monte Carlo-Based Algorithms

3.1. Bayes Filtering

In a state-space model, the transition function and the measurement model can be expressed by

$$x_k = f_k(x_{k-1}, \epsilon_k), \tag{10}$$

$$y_k = h_k(x_k, e_k), \tag{11}$$

where y_k is the measurement vector at epoch k; $f_k()$ is the transition function; $h_k()$ is the measurement function; ϵ_k and e_k are the state noise and the measurement noise, respectively. This model indicates a first-order Markov process because the estimated state vector is only related to the states of the previous epoch $k-1$, but not to other states before.

Considering the Chapman–Kolmogorov equation [31],

$$p(x_k|y_{1:k-1}) = \int p(x_k|x_{k-1}) p(x_{k-1}|y_{1:k-1}) dx_{k-1}, \tag{12}$$

the posterior density $p(x_k|y_{1:k})$ can be estimated according to the Bayes's theorem by

$$p(x_k|y_{1:k}) = p(y_k|x_k) p(x_k|y_{1:k-1}) / p(y_k|y_{1:k-1}). \tag{13}$$

The expectation of x can, hence, be expressed by

$$\hat{x} = \int x p(x_k|y_{1:k}) dx. \tag{14}$$

Combining Equations (12) and (13), the estimates of x on each epoch can be calculated theoretically. The optimal analytical expression for $p(x_k|y_{1:k})$ can be derived for a linear Gauss–Markov model as a Kalman filter but cannot be obtained for most cases. Fortunately, the suboptimal solutions by the Monte Carlo method are usually available.

3.2. Importance Sampling

In Monte Carlo methods, the probability density function (PDF) $p(x_k|y_{1:k})$ is represented by N samples $\{x^i\}_{i=1}^N$; therefore,

$$p(x_k|y_{1:k}) \approx \frac{1}{N} \sum_{i=1}^N \delta(x - x^i) \Rightarrow \hat{x} \approx E(\{x^i\}_{i=1}^N), \tag{15}$$

where $\delta()$ is the Dirac delta function.

The posterior density is not precisely known at the beginning of epoch k. Assuming a prior PDF $q(x)$ is known from which the samples can be generated and $p(x)$ is the posterior density to be estimated, after the samples are generated, they can be used to load the information provided by the measurements at epoch k and obtain more precisely posterior density. The expectation of the unknown state vector can be calculated by

$$\hat{x} = \int xp(x)dx = \int xp(x)/q(x)q(x)dx = \int x(p(x)/q(x))q(x)dx = \int x\,w(x)\,q(x)dx, \quad (16)$$

where $w(x) = p(x)/q(x)$; $q(x)$ is the importance density function.

As the PDFs are represented via the Monte Carlo method and according to the Bayes's theorem, the expectation can be written as

$$\hat{x} \approx \frac{1}{N}\sum_{i=1}^{N} w^i x^i, \quad (17)$$

where w^i of x_k is derived by $\widetilde{w}(x_k) = \sum_{i=1}^{N} w_{k-1}^i p(y_k|x_k)(p(x_k|x_{k-1}^i)/q(x_k|x_{k-1}^i, y_k))$.

3.3. Sequential Monte Carlo Algorithm

A sequential importance sampling (SIS) procedure can be implemented to get estimates of x. Firstly, sample set $\{x_0^i\}_{i=1}^{N}$ is generated with equal weights according to the initial distribution $q(x_0)$. Then, the samples are reweighted according to the likelihood of the measurements and the estimates $\hat{x}$ are calculated. Afterward, the samples are transited to the next epoch. In practice, the SIS quickly degenerates during the filtering as more and more particles get negligible weights. Fortunately, the degeneracy problem can be solved by resampling which duplicates the samples with large weights and deletes the samples with small weights.

The resampling step is implemented as follows: (a) the numerical CDF $\{W_k^i\}_{i=1}^{N}$ of x is constructed with $W_k^i = \sum_{j=1}^{i} w_k^j$; (b)CDF $\{u_i\}_{i=1}^{N}$ is generated by $u_i = ((i-1)+\widetilde{u})/N$, where $\widetilde{u}$ is a random value over interval [0, 1); (c) m = 1; for each i = 1, ..., N, if $\widetilde{w}_k^m < u_i$, x_k^m is deleted by setting $m = m+1$; otherwise, x_k^m is duplicated by setting $\overline{x}_k^j = x_k^m$; (d) the new sample set $\{\overline{x}_k^j\}_{j=1}^{N}$ is assigned equal weights.

It is not necessary to resample the samples each epoch, and a condition can be set by comparing the effective number with a threshold. This resampling procedure adequately solves the degeneracy problem in SIS in practice. Considering the resampling step, the SMC procedure can be implemented as Algorithm 1.

Algorithm 1: Sequential Monte Carlo (SMC)

(a) Initialization	Generate samples $\{x_0^i\}_{i=1}^{N}$, with $x_0^i \sim q(x_0)$.
(2) Update	Update the weights according to likelihood function $p(y_k|x_k^i)$ of measurements with $\overline{w}_k^i = w_{k-1}^i p(y_k|x_k^i)$.
	Normalize the weights by $w_k^i = \overline{w}_k^i / \sum_{j=1}^{N} \overline{w}_k^j$
	Calculate the estimated value and variance by $\hat{x}_k \approx \sum_{i=1}^{N} x_k^i w_k^i$ and $\text{var}(\hat{x}_k) \approx \sum_{i=1}^{N}(x_k^i - \hat{x}_k)(x_k^i - \hat{x}_k)^T w_k^i$, respectively.
(c) Resampling	Implement resampling if $N_{eff} < N_{th}$,
	where N_{eff} is the effective number of samples which is calculated by $N_{eff} = 1/\sum_{i=0}^{N}(w_k^i)^2$
	and N_{th} is a threshold which can be set to the value of $2/3N$.
(d) Prediction	Draw new samples $\{x_k^i\}_{i=1}^{N}$, by $x_k^i = f(x_{k-1}^i) + v_k$.
	Repeat steps (b) to (d) for the next epoch $k + 1$.

3.4. Sequential Quasi-Monte Carlo Algorithm

The pseudo random numbers usually used in the Monte Carlo method encounter possible gaps and clusters in the sampling fields. This can be avoided by a QMC method which replaces the random sequences with low-discrepancy sequences such as the Sobol sequence, Halten sequence, and so on [32,33].

The QMC sequence is a deterministic sequence and is uniformly distributed over $[0,1]^d$. Let $\{u^i\}_{i=1}^N$ be a sequence of vectors in $[0,1]^d$ and $= [0,x]$ be a subinterval of $[0,1]^d$. is the sub sequence of $\{u^i\}_{i=1}^N$ first belonging to . The discrepancy of the sequence is defined by

$$D^*\left(\{u^i\}_{i=1}^N\right) = \sup_{x \in [0,1]^d} \left| \frac{\# \text{ of }}{N} - \prod_{j=1}^d x^j \right|, \qquad (18)$$

where sup refers to the supremum. A sequence with small discrepancy defined by the above formula is named a low-discrepancy sequence. Koksma–Hlawka theorem indicates that the approximation error of a real function represented by discrete numbers, i.e., the left side of Equation (19), is bounded by the product of two independent factors, the variation of the real function, and the discrepancy of the discrete numbers. Therefore, we have the following inequality:

$$\left| \frac{1}{N} \sum_{i=1}^N f(u^i) - \int_{[0,1]^d} f(u)du \right| \leq V(f) D^*\left(\{u^i\}_{i=1}^N\right), \qquad (19)$$

where $V(f)$ is the variation of f in the sense of Hardy and Krause [34]. A low-discrepancy sequence has $D^*\left(\{u^i\}_{i=1}^N\right) = O(N^{-1}(\ln^d N))$. This indicates that the QMC estimate for numerical integration has a probabilistic error bound of $O(N^{-1}(\ln^d N))$, which is better than the error bound of MC estimate $O(N^{-1/2})$. This can improve the convergence and enables efficient computing.

It is difficult to analyze the accuracy of the approximation by QMC in practice as the points are regular. Therefore, randomized QMC (RQMC) can be used so that every element of the sequence is uniformly distributed over the unit cub but still has a low-discrepancy property [35–37]. Figure 1 shows the first 200 samples of the sequences for RQMC sampling, pseudo-random sampling, and the corresponding Gaussian sampling with a Sobol sequence. The SQMC algorithm is presented in Algorithm 2.

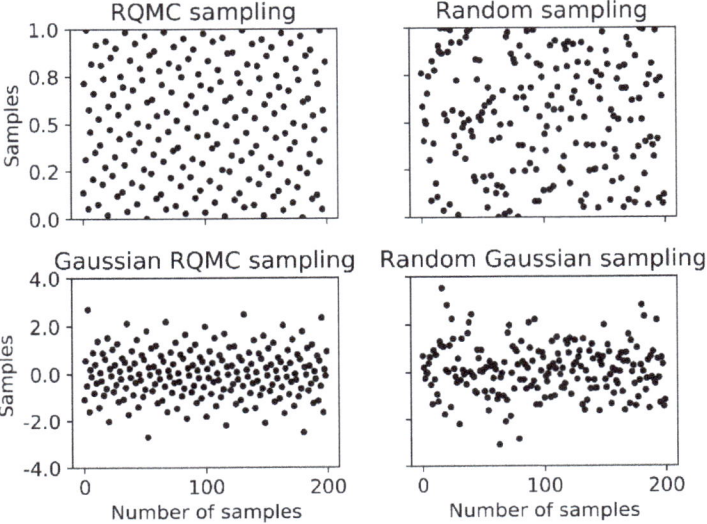

Figure 1. The 200 points investigated for the randomized quasi-Monte Carlo (RQMC) sampling, pseudo-random sampling, and the corresponding Gaussian sampling with a Sobol sequence.

	Algorithm 2: SQMC
(a) Initialization	Generate a QMC or RQMC points $\{u^i\}_{i=1}^{N}$ where $u^i \in [0,1]^d$; generate $\{x_0^i\}_{i=1}^{N}$ according to $x_0^i = q^{-1}(u^i)$, where $q(x)$ is the prior density of x.
(b) Update	Update the weights according to likelihood function $p(y_k|x_k^i)$ of measurements with $\overline{w}_k^i = w_{k-1}^i p(y_k|x_k^i)$.
	Normalize the weights by $w_k^i = \overline{w}_k^i / \sum_{j=1}^{N} \overline{w}_{k'}^j$. Calculate the estimated value and variance by $\hat{x}_k \approx \sum_{i=1}^{N} x_k^i w_k^i$ and $\mathrm{var}(\hat{x}_k) \approx \sum_{i=1}^{N} (x_k^i - \hat{x}_k)(x_k^i - \hat{x}_k)^T w_k^i$, respectively.
(c) Resample	Resample if $N_{eff} < N_{th}$, where N_{eff} is the effective number of samples which is calculated by $N_{eff} = 1/\sum_{i=0}^{N}(w_k^i)^2$ and N_{th} is a threshold which can be set to the value of $2/3N$.
(d) prediction	Generate a QMC or RQMC points $\{u^i\}_{i=1}^{N}$ where $u^i \in [0,1]^d$, draw new samples $\{x_k^i\}_{i=1}^{N}$ according to $p(x_k|y_{1:k-1})$ with $\{u^i\}_{i=1}^{N}$.
	Repeat steps (b) and (c) for the following epochs.

4. SQMC-Based Algorithm for Phase Bias Estimation

The ratio for integer ambiguity validation in step 3 of Section 2 reflects the closeness of the float ambiguity vector to the integer ambiguity vector and, thus, shows the quality of the ambiguity fixing performance. If the ambiguity is successfully fixed to the integer ambiguities with high probability, the phase bias can be precisely derived. Although the searching and validation step cannot be linearized to satisfy the conditions for using the linear least-squares methods, we can count on the Monte Carlo-based method to develop algorithms for precise phase bias estimation.

The ratio value used in the ambiguity validation reflects the quality of the integer ambiguity fixing performance as used in the ambiguity validation. If b_k is the correct ambiguity vector and x_k represents the phase bias parameters at epoch k, we can have the assumption that the conditional probability density $p(b_k|x_k)$ has the proportional relationship $p(b_k|x_k) \propto ratio(x_k)$, and simply let

$$p(b_k|x_k^i) = ratio(x_k^i) / \sum_{i=1}^{N} ratio(x_k^i). \tag{20}$$

The PDF $p(b_k|x_k)$ is then used as the likelihood function in the Monte Carlo-based estimation method to update the weights. This is expressed as

$$\overline{w}_k^i = w_{k-1}^i p(b_k|x_k^i) = w_{k-1}^i ratio(x_k^i) / \sum_{i=1}^{N} ratio(x_k^i). \tag{21}$$

This designed likelihood function works for the estimation of the phase biases which affect the ratio values in ambiguity fixing.

The following procedure is implemented to calculate $ratio(x_k^i)$ at epoch k for each element in sample set $\{x_k^i\}_{i=1}^{N}$: (a) x_k^i is used as known bias values to calibrate the measurement model by Equation (4) and solve the equation set to get float SD ambiguities and the corresponding VC matrix; (b) the DD ambiguities and the VC matrix are calculated, and the integer ambiguities are fixed using the LAMBDA method; (c) $ratio(x_k^i)$ is calculated using Equation (7).

Moreover, the phase bias can be regarded as constant between epochs, and the transition function which transports samples from epoch $k-1$ to epoch k is

$$x_k^i = x_{k-1}^i + v^i, \tag{22}$$

where v is the normal distributed noise with each element $v \sim N(0, \sigma)$.

The flowchart of the SQMC procedure for phase bias estimation is plotted in Figure 2, and the corresponding algorithm is presented as Algorithm 3.

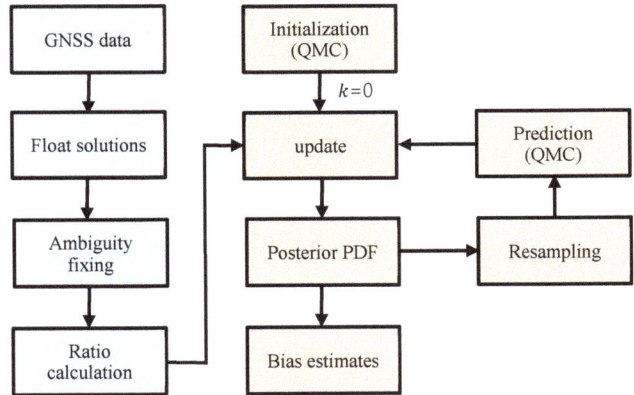

Figure 2. Flowchart of the global navigation satellite system (GNSS) phase bias estimation. The blue boxes refer to the GNSS data processing steps and the yellow boxes are related to SQMC.

Algorithm 3: GNSS phase bias estimation (SQMC)	
Initialization	Generate a QMC or RQMC points $\{u^i\}_{i=1}^N$ where $u^i \in [0,1]^d$; generate $\{x_0^i\}_{i=1}^N$ according to $x_0^i = q^{-1}(u^i)$, where $q(x)$ is the prior density of x.
Update	Update the weights according to likelihood function $p(y_k|x_k^i)$ of measurements with $\overline{w}_k^i = w_{k-1}^i p(b_k|x_k^i)$.
	Normalize the weights by $w_k^i = \overline{w}_k^i / \sum_{j=1}^N \overline{w}_k^j$, Calculate the estimated value and variance by $\hat{x}_k \approx \sum_{i=1}^N x_k^i w_k^i$ and $\text{var}(\hat{x}_k) \approx \sum_{i=1}^N (x_k^i - \hat{x}_k)(x_k^i - \hat{x}_k)^T w_k^i$, respectively.
Resample	Resample if $N_{eff} < N_{th}$,
	where N_{eff} is the effective number of samples which is calculated by $N_{eff} = 1/\sum_{i=0}^N (w_k^i)^2$ and N_{th} is a threshold which can be set to the value of $\frac{2}{3} N$.
prediction	Generate a QMC or RQMC points $\{u^i\}_{i=1}^N$ where $u^i \in [0,1]^d$; draw new samples $\{x_k^i\}_{i=1}^N$ with noise of $N^{-1}(\{u^i\}_{i=1}^N)$.
	Repeat steps 2 and 3 for the following epochs.

Algorithm 3 combines the QMC method and the GNSS ambiguity fixing procedures together to estimate the GNSS phase bias. The low-discrepancy sequences of QMC are included for variance reduction. Section 5 shows the applications of the approach with practical GNSS data.

5. Experiments with Practical GNSS Data

The GLONASS phase IFB estimation of baseline GOP6_GOP7 in networks of international GNSS service (IGS) (ftp://ftp.cddis.eosdis.nasa.gov/pub/gnss) was taken as an example to demonstrate the variance reduction by QMC. The baseline was in Europe with the location in Figure 3, and the two GNSS stations were equipped with LEICA GRX1200+GNSS 9.20 and TRIMBLE NETR9 5.01 receivers, respectively. The measurement data were collected at GPS time (GPST) 9:00–10:00 a.m. on day of year (DOY) 180 of 2018 with an epoch interval of 30 seconds. Six GLONASS satellites were observed during the time span, and the satellite slot numbers are shown at the beginning of the satellite trajectories in Figure 4. The baseline had a post-processed GLONASS phase IFB around −29.5 mm/frequency number (FN) which can be regarded as the true values of both L1 and L2 frequencies.

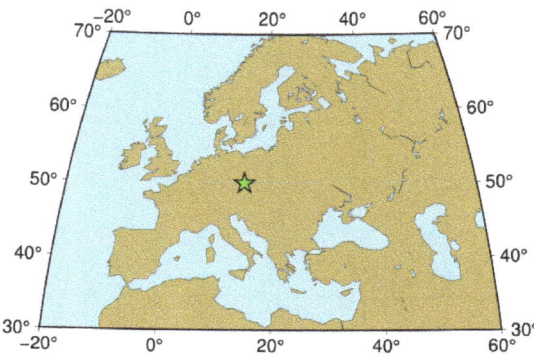

Figure 3. Location of the GNSS baseline GOP6_GOP7 in Europe.

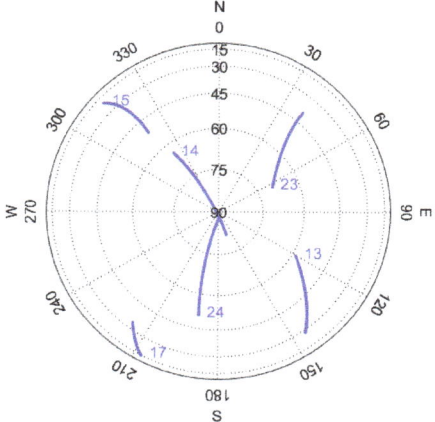

Figure 4. Observed Russia's Global Navigation Satellite System (GLONASS) satellites shared by the two stations of baseline GOP6_GOP7 at GPST 9:00–10:00 a.m. on day of year (DOY) 180 of 2018. The satellite slot numbers in blue color are marked at the beginning of the satellite trajectories.

Both the code and carrier phase measurements of frequency L1 and L2 were used to form Equation (1) in the experiment. Only one IFB parameter was included because the IFB values for both frequency L1 and L2 were regarded as the same. Algorithm 3 for SQMC-based phase-bias estimation was implemented, and the IFB estimate was derived at each epoch. Furthermore, the IFB estimates were also calculated using the SMC-based approach for comparison. When the bias was estimated many times, the estimates of each time were different because the RQMC sequence was used in the SQMC-based procedure, and the pseudo-random sequence was used in the SMC-based procedure. The standard deviation (SD) of the estimates was calculated to evaluate the performance.

Firstly, the IFB was estimated 1000 times with the SMC-based approach. The sigma of the transition noise in Equation (22) was set to 1 mm/FN, and the sample number was fixed to a value of 100. The 1000 estimates of IFB for the first 60 epochs are drawn in Figure 5 as yellow lines. Afterward, the SQMC-based approach with a Sobol sequence for IFB estimation was implemented. The SQMC strategy also had a sigma of the transition noise as 1 mm/FN and the number of samples as 100. The IFB was also estimated 1000 times, and the results for the first 60 epochs are plotted in Figure 5 as blue lines. The SDs of the 1000 estimates of SMC and SQMC approaches are calculated and presented in Figure 5 as a yellow line and blue line, respectively.

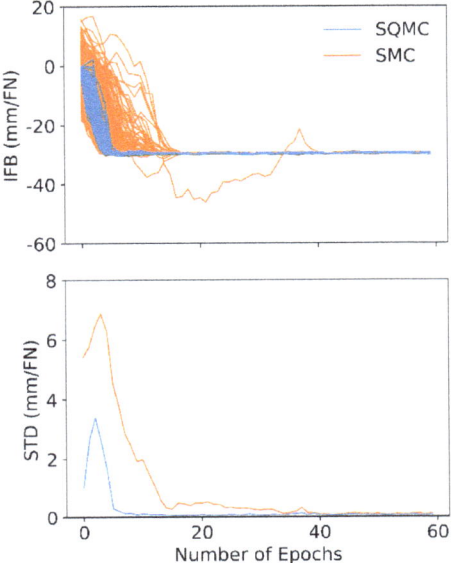

Figure 5. Inter-frequency bias (IFB) estimates with 1000 iterations and their SDs based on SQMC-based and SMC-based algorithms with data of baseline GOP6_GOP7 collected at GPST 9:00–9:30 a.m. on DOY 180 of 2018.

Obviously, both approaches could successfully estimate the IFB values. After convergence, the estimates of the two algorithms were very similar, such as the results after epoch 40th. However, the estimates of the SQMC-based algorithm converged faster than those of the SMC approach, and the corresponding SD was much lower at the beginning. In the worst case of the 1000 estimates, the results converged, i.e., became close to the true value with a difference smaller than 1.5 mm/FN at the seventh epoch for SQMC and at the 40th epoch for SMC.

The variance of the estimates for the SQMC-based algorithm with variation in the sample numbers and the transition noise was also analyzed.

Firstly, the phase IFB was estimated 100 times using the SMC-based and SQMC-based algorithms, separately, with the number of particles varied from 30 to 200. The sigma of transition noise was fixed to 1 mm/FN for each estimation. The SDs of the 100 estimates at epochs 1, 2, 5, and 10 are presented in Figure 6, where we can see that the SD decreased for both SMC and SQMC as the number of particles increased. The SD for SQMC had much smaller values compared with SQMC. This indicates that the SQMC-based algorithm could achieve estimates with smaller SD than the SMC-based algorithm using even smaller sample numbers. This is very meaningful in GNSS data processing, because the main time-consuming step is the ambiguity fixing procedure in step 2 in Section 2 for each sample. Fewer samples result in a lighter computation load.

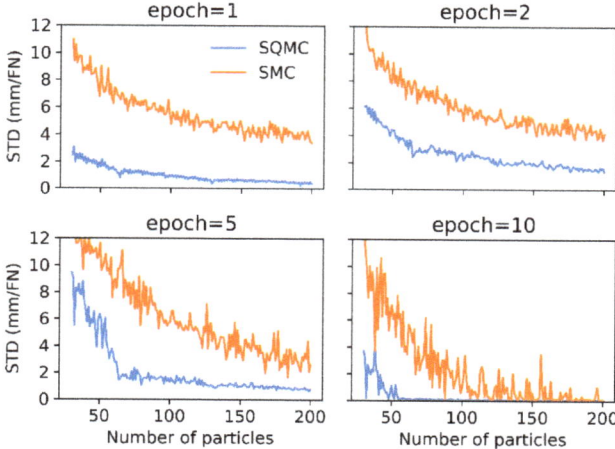

Figure 6. SDs of the 100 estimates based on SQMC-based and SMC-based algorithms, separately, for different number of particles from 30 to 200 at epochs 1, 2, 5, and 10. The data are for baseline GOP6_GOP7 collected at GPST 9:00–9:30 a.m. on DOY 180 of 2018.

Then, the effects of the transition noise were evaluated. The phase IFB was estimated another 100 times with the sigma of the transition noise from 10^{-6} to 10^{-2} m/FN, and the number of the samples was fixed to 100. The SDs of the 100 estimates were calculated at each epoch and the values at epochs 1, 2, 5, and 10 are plotted in Figure 7. Obviously, the SDs of the SQMC-based algorithm were much smaller than those of the SMC-based algorithm at all the four epochs. The SDs at epoch 10 showed a curve near 10^{-3} m/FN and were larger than the STDs corresponding to other nearby sigma values. This indicates that the transition noise level in the transition model needs to be set carefully. Too high a transition noise will increase the SDs; however, if the sigma is too small, the samples cannot evolve to the proper field and the prior density cannot be well represented by the samples, also leading to large STD values.

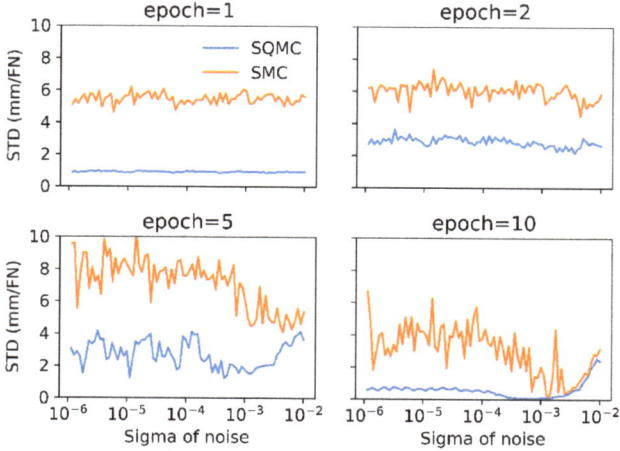

Figure 7. SD of the 100 estimates based on SQMC-based and SMC-based algorithms, separately, for different transition noise with sigma from 10^{-6} to 10^{-2} m/FN at epochs 1, 2, 5, and 10. The data are for baseline GOP6_GOP7 collected at GPST 9:00–9:30 a.m. on DOY 180 of 2018.

6. Conclusions

This article presented the problem of solving the mathematical models in GNSS phase bias estimation, which is essential for fast and precise GNSS data processing; it then developed a fast and efficient algorithm by combining the GNSS data processing procedure and the SQMC method.

The proposed SQMC-based GNSS phase bias estimation algorithm introduces the QMC technique to the SMC-based algorithm for variance reduction. The random sequences in SMC approach are replaced by low-discrepancy sequences in the sampling of the steps of initialization and prediction. We performed the experiments of phase IFB estimation for GLONASS data processing with practical data. The results show that the GNSS phase bias estimates have much smaller variance based on the low-discrepancy sequence. The estimation converges faster, and the results are more precise even with a much smaller number of samples. This can largely save the convergence time and the computation load of GNSS PNT services.

In addition, this study introduces the difficulty in GNSS data processing to the mathematic community, and it has the potential to boost new numerical algorithms for GNSS research and applications.

Author Contributions: Conceptualization, Y.T., M.G., and F.N.; methodology, Y.T., M.G., and F.N.; writing—original draft preparation, Y.T.; writing—review and editing, M.G. and F.N. All authors read and agreed to the published version of the manuscript.

Funding: This research was funded by the National Natural Science Foundation of China (No. 41804022) and the Fundamental Research Funds for the Central Universities (No. 2682018CX33).

Acknowledgments: The authors thank IGS MGEX for providing GNSS data.

Conflicts of Interest: The authors declare no conflicts of interest.

Abbreviations

The following abbreviations are used in this manuscript:

BDS	BeiDou Navigation Satellite System
DOY	Day of year
GLONASS	GLObal NAvigation Satellite System
GNSS	Global navigation satellite system
GPS	Global positioning system
IFB	Inter-frequency bias
IGS	International GNSS Service
ISB	Inter-system bias
LAMBDA	Least-squares ambiguity decorrelation adjustment
MC	Monte Carlo method
PDF	Probability density function
PNT	Positioning, navigation, and timing
QMC	Quasi-Monte Carlo method
SIS	Sequential importance sampling
SMC	Sequential Monte Carlo method
SQMC	Sequential quasi-Monte Carlo method
STD	Standard deviation
VC	Variance–covariance

References

1. Steigenberger, P.; Montenbruck, O. Galileo status: Orbits, clocks, and positioning. *GPS Solut.* **2017**, *21*, 319–331. [CrossRef]
2. Yang, Y.X.; Gao, W.; Guo, S.; Mao, Y.; Yang, Y.F. Introduction to BeiDou-3 navigation satellite system. *Navigation* **2019**, *66*, 7–18. [CrossRef]

3. Montenbruck, O.; Steigenberger, P.; Prange, L.; Deng, Z.; Zhao, Q.; Perosanz, F.; Romero, I.; Noll, C.; Stürze, A.; Weber, G.; et al. The Multi-GNSS Experiment (MGEX) of the International GNSS Service (IGS)–achievements, prospects and challenges. *Adv. Space Res.* **2017**, *59*, 1671–1697. [CrossRef]
4. Khodabandeh, A.; Teunissen, P. PPP-RTK and inter-system biases: The ISB look-up table as a means to support multi-system PPP-RTK. *J. Geod.* **2016**, *90*, 837–851. [CrossRef]
5. Prange, L.; Orliac, E.; Dach, R.; Arnold, D.; Beutler, G.; Schaer, S.; Jäggi, A. CODE's five-system orbit and clock solution—The challenges of multi-GNSS data analysis. *J. Geod.* **2017**, *91*, 345–360. [CrossRef]
6. Teunissen, P. The least-squares ambiguity decorrelation adjustment: A method for fast GPS integer ambiguity estimation. *J. Geod.* **1995**, *70*, 65–82. [CrossRef]
7. Hofmann-Wellenhof, B.; Lichtenegger, H.; Wasle, E. *GNSS–Global Navigation Satellite Systems: GPS, GLONASS, Galileo, and More*; Springer: Vienna, Austria, 2008.
8. Chang, X.; Yang, X.; Zhou, T. MLAMBDA: A modified LAMBDA method for integer least-squares estimation. *J. Geod.* **2005**, *79*, 552–565. [CrossRef]
9. Odijk, D.; Nadarajah, N.; Zaminpardazm, S.; Teunissen, P. GPS, Galileo, QZSS and IRNSS differential ISBs: Estimation and application. *GPS Solut.* **2016**, *21*, 439–450. [CrossRef]
10. Paziewski, J.; Wielgosz, P. Accounting for Galileo-GPS inter-system biases in precise satellite positioning. *J. Geod.* **2015**, *89*, 81–93. [CrossRef]
11. Odijk, D.; Zhang, B.; Khodabandeh, A.; Odolinski, R.; Teunissen, P. On the estimability of parameters in undifferenced, uncombined GNSS network and PPP-RTK user models by means of S-system theory. *J. Geod.* **2016**, *90*, 15–44. [CrossRef]
12. Tian, Y.; Ge, M.; Neitzel, F.; Zhu, J. Particle filter-based estimation of inter-system phase bias for real-time integer ambiguity resolution. *GPS Solut.* **2017**, *21*, 949–961. [CrossRef]
13. Tian, Y.; Ge, M.; Neitzel, F. Particle filter-based estimation of inter-frequency phase bias for real-time GLONASS integer ambiguity resolution. *J. Geod.* **2015**, *89*, 1145–1158. [CrossRef]
14. Sun, X.; Chao, H.; Pei, C. Instantaneous GNSS attitude determination: A Monte Carlo sampling approach. *Acta Astronaut.* **2017**, *133*, 24–29. [CrossRef]
15. Cheng, C.; Tourneret, J.; Pan, Q.; Calmettes, V. Detecting, estimating and correcting multipath biases affecting GNSS signals using a marginalized likelihood ratio-based method. *Signal Process.* **2016**, *118*, 221–234. [CrossRef]
16. Gordon, N.; Salmond, D.; Smith, A.F.M. Novel approach to nonlinear and non-Gaussian Bayesian state estimation. *Proc. Inst. Elect. Eng. F* **1993**, *140*, 107–113. [CrossRef]
17. Arulampalam, S.; Maskell, S.; Gordon, N.; Clapp, T. A tutorial on particle filters for online nonlinear/non Gaussian Bayesian tracking. *IEEE Trans. Signal Process.* **2002**, *50*, 174–188. [CrossRef]
18. Flury, T.; Shephard, N. Bayesian inference based only on simulated likelihood: Particle filter analysis of dynamic economic models. *Economet Theor.* **2011**, *27*, 933–956. [CrossRef]
19. Kantas, N.; Doucet, A.; Singh, S.S.; Maciejowski, J.; Chopin, N. On particle methods for parameter estimation in state-space models. *Stat. Sci.* **2015**, *30*, 328–351. [CrossRef]
20. Doucet, A.; Smith, A.; Freitas, N.; Gordon, N. *Sequential Monte Carlo Methods in Practice*; Springer: New York, NY, USA, 2001.
21. Fearnhead, P.; Kuensch, H.R. Particle Filters and Data Assimilation. *Annu. Rev. Stat. Appl.* **2018**, *5*, 421–449. [CrossRef]
22. Fearnhead, P. Using random quasi-Monte-Carlo within particle filters, with application to financial time series. *J. Comput. Graph. Stat.* **2005**, *14*, 751–769. [CrossRef]
23. Guo, D.; Wang, X. Quasi-Monte Carlo filtering in nonlinear dynamic systems. *IEEE Trans. Signal Process.* **2006**, *54*, 2087–2098.
24. Chowdhury Roy, S.; Roy, D.; Vasu, R.M. Variance-reduced particle filters for structural system identification problems. *J. Eng. Mech.* **2013**, *139*, 210–218. [CrossRef]
25. Gerber, M.; Chopin, N. Sequential Quasi Monte Carlo. *J. R. Stat. Soc.* **2015**, *77*, 509–579. [CrossRef]
26. Strasser, S.; Mayer-Gürr, T.; Zehentner, N. Processing of GNSS constellations and ground station networks using the raw observation approach. *J. Geod.* **2019**, *93*, 1045–1057. [CrossRef]
27. Wanninger, L. Carrier-phase inter-frequency biases of GLONASS receivers. *J. Geod.* **2012**, *86*, 139–148. [CrossRef]

28. Ge, M.; Gendt, G.; Rothacher, M.; Shi, C.; Liu, J. Resolution of GPS carrier-phase ambiguities in precise point positioning (PPP) with daily observations. *J. Geod.* **2008**, *82*, 38. [CrossRef]
29. Verhagen, S.; Teunissen, P. The ratio test for future GNSS ambiguity resolution. *GPS Solut.* **2013**, *17*, 535–548. [CrossRef]
30. Euler, H.; Schaffrin, B. On a measure of discernibility between different ambiguity solutions in the static kinematic GPS-mode. In *Kinematic Systems in Geodesy, Surveying, and Remote Sensing*; Schwarz, K., Lachapelle, G., Eds.; Springer: Berlin/Heidelberg, Germany; New York, NY, USA, 1991; pp. 285–295.
31. Hughes, B. *Random Walks and Random Environments. Volume 1: Random Walks*; Clarendon Press: Oxford, UK, 1995.
32. Owen, A.B. Scrambling Sobol'and Niederreiter–Xing Points. *J. Complex.* **1998**, *14*, 466–489. [CrossRef]
33. L'Ecuyer, P.; Lemieux, C. Recent Advances in Randomized Quasi-Monte Carlo Methods. In *Modeling Uncertainty: An Examination of Stochastic Theory, Methods, and Applications*; Springer: New York, NY, USA, 2002; pp. 417–474.
34. Tuffin, Bruno. On the use of low discrepancy sequences in Monte Carlo methods. *Monte Carlo Methods Appl.* **1996**, *2*, 295–320.
35. L'Ecuyer, P. Randomized quasi-Monte Carlo: An introduction for practitioners. In Proceedings of the International Conference on Monte Carlo and Quasi-Monte Carlo Methods in Scientific Computing, Stanford, CA, USA, 14–19 August 2016.
36. Vermaak, J.; Godsill, S.J.; Perez, P. Monte Carlo filtering for multi target tracking and data association. *IEEE Trans. Aerosp. Electron. Syst.* **2005**, *41*, 309–332. [CrossRef]
37. Hong, H.S.; Hickernell, F.J. Algorithm 823: Implementing Scrambled Digital Sequences. *ACM Trans. Math. Softw.* **2003**, *29*, 95–109. [CrossRef]

© 2020 by the authors. Licensee MDPI, Basel, Switzerland. This article is an open access article distributed under the terms and conditions of the Creative Commons Attribution (CC BY) license (http://creativecommons.org/licenses/by/4.0/).

Article

Automatic Calibration of Process Noise Matrix and Measurement Noise Covariance for Multi-GNSS Precise Point Positioning

Xinggang Zhang [1,2,3], **Pan Li** [3,*], **Rui Tu** [1,2], **Xiaochun Lu** [1,2], **Maorong Ge** [3,4] **and Harald Schuh** [3,4]

1. National Time Service Center, Chinese Academy of Sciences, Shu Yuan Road, Xi'an 710600, China; zxg@ntsc.ac.cn (X.Z.); turui-2014@126.com (R.T.); luxc@ntsc.ac.cn (X.L.)
2. Key Laboratory of Precision Navigation Positioning and Timing Technology, Chinese Academy of Sciences, Xi'an 710600, China
3. German Research Centre for Geosciences (GFZ), 14473 Potsdam, Germany; maorong.ge@gfz-potsdam.de (M.G.); schuh@gfz-potsdam.de (H.S.)
4. Institute of Geodesy and Geoinformation Science, Technische Universität Berlin, 10623 Berlin, Germany
* Correspondence: panli@gfz-potsdam.de

Received: 28 January 2020; Accepted: 12 March 2020; Published: 2 April 2020

Abstract: The Expectation-Maximization algorithm is adapted to the extended Kalman filter to multiple GNSS Precise Point Positioning (PPP), named EM-PPP. EM-PPP considers better the compatibility of multiple GNSS data processing and characteristics of receiver motion, targeting to calibrate the process noise matrix Q_t and observation matrix R_t, having influence on PPP convergence time and precision, with other parameters. It is possibly a feasible way to estimate a large number of parameters to a certain extent for its simplicity and easy implementation. We also compare EM-algorithm with other methods like least-squares (co)variance component estimation (LS-VCE), maximum likelihood estimation (MLE), showing that EM-algorithm from restricted maximum likelihood (REML) will be identical to LS-VCE if certain weight matrix is chosen for LS-VCE. To assess the performance of the approach, daily observations from a network of 14 globally distributed International GNSS Service (IGS) multi-GNSS stations were processed using ionosphere-free combinations. The stations were assumed to be in kinematic motion with initial random walk noise of 1 mm every 30 s. The initial standard deviations for ionosphere-free code and carrier phase measurements are set to 3 m and 0.03 m, respectively, independent of the satellite elevation angle. It is shown that the calibrated R_t agrees well with observation residuals, reflecting effects of the accuracy of different satellite precise product and receiver-satellite geometry variations, and effectively resisting outliers. The calibrated Q_t converges to its true value after about 50 iterations in our case. A kinematic test was also performed to derive 1 Hz GPS displacements, showing the RMSs and STDs w.r.t. real-time kinematic (RTK) are improved and the proper Q_t is found out at the same time. According to our analysis despite the criticism that EM-PPP is very time-consuming because a large number of parameters are calculated and the first-order convergence of EM-algorithm, it is a numerically stable and simple approach to consider the temporal nature of state-space model of PPP, in particular when Q_t and R_t are not known well, its performance without fixing ambiguities can even parallel to traditional PPP-RTK.

Keywords: EM-algorithm; multi-GNSS; PPP; process noise; observation covariance matrix; extended Kalman filter; machine learning

1. Introduction

Since Precise Point Positioning (PPP) emerged [1,2], people are primarily focusing on improving precise orbit and clock products, developing new algorithms to solve for ambiguities, to accelerate its convergence and expand its applications such as PPP-real-time-kinematic (PPP-RTK), triple frequency PPP [3], ionosphere-constraint PPP and low-cost receiver PPP [4–10].

Generally, PPP can be realized by the least-squares method (including sequential least-squares) or extended Kalman filter. The least-squares method is for static state estimation and thus does not reflect varying user dynamics. To work the same as Kalman filter, the process noise matrix is added to the gain matrix of the sequential least-squares method to adjust receiver clock behavior and atmospheric activity and so on, which is named as a sequential filter [2]. Hence, in the following paper, the authors will only consider the Kalman filter for PPP data processing.

Although both the process noise Q_t and observation covariance matrix R_t are the key to Kalman filter, limited attention is paid to the fundamental problem for multi-GNSS PPP. Q_t and R_t must be consistent with state dynamics and measurement accuracy, respectively. For example, if the value of Q_t is too small, the estimated state will lose its minimum mean squared error property, and if the value of Q_t is too large with respect to the correct one, the estimated state will oscillate around the true value. Moreover, because of ground deformation and specific surroundings, Q_t should not be kept fixed to calculate the optimal estimates. In other words, Q_t should evolve with time and a proper Q_t will shorten the PPP convergence time. If Q_t is improper, it may damage PPP convergence sometimes.

As for GNSS, R_t not only depends on the measurement accuracy, elevation of the satellite, orbit and clock error, atmospheric delay error, multipath, missing data, etc. but possibly deteriorate after a lapse of a period. What is more, the assumption, frequently used in geodesy, that different types of measurements have a fixed error ratio is not always true, because the ratio is closely linked to receivers and antenna types, and to the performance of satellite system itself. For example, while fusing multiple GNSS, the weight of measurements of GPS is intended to be higher, other GNSS systems to be lower. It is not easy to give a prior accurate ratio.

Generally, four methods are often applicable to calibrate Q_t and R_t. The first one is based on the innovation property of the Kalman filter, in which a moving-window recursive way is used to identify Q_t and R_t [11–14]. However, none of them can maintain the positive semi-definiteness of the estimated covariances. To solve this problem, Odelson developed the autocovariance least-squares method for estimating covariances using a lagged autocovariance function [15]. This kind of least-squares method depends on the user-defined autocovariance function.

The second scheme to recognize Q_t and R_t is the multiple model adaptive estimation (MMAE) [13,16]. MMAE runs a bank of Kalman filter in parallel, every one of them is driven by its pair of Q_t and R_t. The final Q_t and R_t are thought of as the weighted sum of the estimates of individual Kalman filter.

In the third scheme, M-estimator is introduced into an adaptive Kalman filter to increase its resistance to outliers, where an adaptive factor α to state error covariance matrix is constructed [17,18]. Yet, choosing a value for α is still very challenging. An improper α will result in biased results.

Another attractive scheme is the least-squares variance component estimation (LS-VCE) [19], which is based on least-squares principles. Similar to restricted maximum likelihood (REML), LS-VCE does not use observations directly but combine observations to exclude any fixed effects. However, LS-VCE needs to define the weight matrix on the user's own and increase its complexity.

In this contribution, a machine learning algorithm, the Expectation-Maximization (EM) algorithm, is developed to the extended Kalman filter to estimate PPP states, $\vec{x}_t$, together with a large number of Q_t and R_t. The EM-algorithm, which can be classified as the first scheme, works in an iterative procedure to locate maximum likelihood estimates of parameters. Its iteration consists of two steps: Expectation and Maximization. In the Expectation step, a function for the expectation of the log-likelihood is computed using the estimates of the current parameters. In the Maximization step, estimates of parameters are updated by maximizing the expected log likelihood function.

On the one hand, the main drawback of the EM-algorithm is that it converges slowly and needs heavy computation. Here the convergence refers to finding maximum likelihood estimator of parameters, not the PPP convergence time. However, for example, its convergence can be accelerated using the Aitken method or conjugate method [20].

On the other hand, it is fairly simple, and has robust convergence and deals conveniently with problems having a lot of parameters. For such problems, it is often the only algorithm to a large extent [19]. It is also capable of finding Kalman parameters even if we have missing data. In addition, it can detect outliers by introducing small weights for large outliers and can even estimate the outliers [21]. In contrast, the outliers are not removed but automatically downweighed in our article, since outliers sometimes take some useful information.

The paper is organized as follows, after clarifying the importance to calibrate Q_t and R_t, the EM algorithm is introduced in the first section. Next, the state space model of PPP is reviewed from a point of machine learning and the methodology to adapt the EM-algorithm for extended Kalman filter for PPP is explained theoretically in detail. Thirdly, we compare the EM-algorithm with other methods, the static and kinematic instances are also given to demonstrate EM-PPP performance to improve the accuracy and reliability of PPP. Finally, the results are analyzed and the conclusion is drawn.

2. State Space Model for PPP

The state-space model allows to process GNSS data in a uniformed form. It is characterized as two equations: the state equation, which comprises a series of vector $\vec{x}_t$, ($1 \leq t \leq N$, N is the number of epochs), and the observation equation. The state $\vec{x}_t$ cannot be observed directly, usually called hidden states, which is driven by hidden process noise. In this article the state-space model is described as the Kalman filter.

2.1. State Equation

The hidden state $\vec{x}_t$ of multi-GNSS PPP Kalman filter involves five types of parameters: three components of receiver coordinates, receiver clock error, system time difference w.r.t. GPS, troposphere zenith wet delay and ambiguities. Using subscript t to denote a specific time epoch, the state at time t evolves from the state at $(t-1)$ according to:

$$\vec{x}_t = \Phi_t \vec{x}_{t-1} + \vec{w}_t \tag{1}$$

where Φ_t is the transition matrix and $\vec{w}_t$ the state process noise, which is assumed to be drawn from a zero-mean multivariate normal distribution, with covariance: $\vec{w}_t \sim \mathcal{N}(0, Q_t)$. Initial condition $\vec{x}_0$ is assumed to be a Gaussian vector with the a priori information $E\{\vec{x}_0\} = \vec{\mu}_0$, $Cov(\vec{x}_0) = P_0$.

The state transition matrix and the process noise matrix in static mode is defined for the position block:

$$\Phi_t = \begin{bmatrix} 1 & 0 & 0 \\ 0 & 1 & 0 \\ 0 & 0 & 1 \end{bmatrix}_t, \quad Q_t = \begin{bmatrix} Q_x & 0 & 0 \\ 0 & Q_y & 0 \\ 0 & 0 & Q_z \end{bmatrix}_t \tag{2}$$

where Q_x, Q_y and Q_z are the random process noise in X, Y and Z direction, respectively.

In addition to position parameters, the velocity parameters are also included in the state vector for our kinematic processing, whose system model for position and velocity block in the extended Kalman filter is given as:

$$\Phi_t = \begin{bmatrix} 1 & \Delta t & 0 & 0 & 0 & 0 \\ 0 & 1 & 0 & 0 & 0 & 0 \\ 0 & 0 & 1 & \Delta t & 0 & 0 \\ 0 & 0 & 0 & 1 & 0 & 0 \\ 0 & 0 & 0 & 0 & 1 & \Delta t \\ 0 & 0 & 0 & 0 & 0 & 1 \end{bmatrix}, \quad Q_t = \begin{bmatrix} \frac{q_x \Delta t^3}{3} & \frac{q_x \Delta t^2}{2} & 0 & 0 & 0 & 0 \\ \frac{q_x \Delta t^2}{2} & q_x \Delta t & 0 & 0 & 0 & 0 \\ 0 & 0 & \frac{q_y \Delta t^3}{3} & \frac{q_y \Delta t^2}{2} & 0 & 0 \\ 0 & 0 & \frac{q_y \Delta t^2}{2} & q_y \Delta t & 0 & 0 \\ 0 & 0 & 0 & 0 & \frac{q_z \Delta t^3}{3} & \frac{q_z \Delta t^2}{2} \\ 0 & 0 & 0 & 0 & \frac{q_z \Delta t^2}{2} & q_z \Delta t \end{bmatrix} \tag{3}$$

The corresponding state vector is $\vec{x}_t = \begin{bmatrix} x & v_x & y & v_y & z & v_z \end{bmatrix}$ for position and velocity block. The process noise matrix Q_t is uniquely determined by q_x, q_y and q_z, which are named as acceleration variance.

2.2. Observation Equation

The observation equation, that we used, is the double frequency ionosphere-free combination for multi-GNSS [22]:

$$PC_r^{i,S} = \rho_r^i + cdt_r^G + T_r^S + m_r^i \cdot ztd_r + \varepsilon_{r,PC}^{i,S} \tag{4}$$

$$LC_r^{i,S} = \rho_r^i + cdt_r^G + T_r^S + m_r^i \cdot ztd_r + \lambda_{LC}^S \cdot N_{r,LC}^S + \varepsilon_{r,LC}^{i,S} \tag{5}$$

where subscript r indicates the receiver, superscript i represents the satellites, superscript S indicates GNSS constellation, following the convention of Rinex3.x (G: GPS, E: GALILEO, R: GLONASS and C: BEIDOU). $PC_r^{i,S}$ and $LC_r^{i,S}$ are the ionosphere-free combinations of code pseudo-range and phase observations (unit: m) respectively, which have already corrected satellite clock, the relativity effect, solid Earth tides, polar tides, ocean tides, phase wind up and a priori troposphere delay using troposphere model [23,24]. ρ_r^i is the geometry distance between receiver and satellite, cdt_r^G is receiver clock (unit: m), the superscript G of cdt_r^G implies that GPS time is selected as the reference time, T_r^S is the system time difference in meters of system S to GPS time. Specifically, for $S = G$, T_r^G is zero. m_r^i is troposphere mapping function and ztd_r is troposphere zenith wet delay, λ_{LC}^S is the wavelength of LC combination corresponding to system S and $N_{r,LC}^S$ is the LC ambiguity. $\varepsilon_{r,PC}^{i,S}$ and $\varepsilon_{r,LC}^{i,S}$ indicate other unmodeled errors or noise.

Equations (4) and (5) are nonlinear, the extended Kalman filter (EKF) can be used for nonlinear state estimation. For easy description, they are rewritten in a general form:

$$\vec{y}_t = h(\vec{x}_t) + \vec{v}_t \tag{6}$$

where $\vec{y}_t = \begin{bmatrix} y_1 & \cdots & y_j & \cdots & y_{k_t} \end{bmatrix}$ is the observation vector, k_t is the number of observations at epoch t, $y_j \in \{PC_r^{i,S}, LC_r^{i,S}\}$, $\vec{v}_t$ is the observation noise satisfying $\vec{v}_t \sim \mathcal{N}(0, R_t)$, R_t is the observation noise covariance matrix at epoch t.

2.3. Kalman Filter

Let $Y_m = \{\vec{y}_1, \ldots, \vec{y}_m\}$ denote all observations from epoch 1 to epoch m, and $\vec{x}_{t|m}$ represent the estimate of $\vec{x}_t$ given observations Y_m, we have predicted state estimate and predicted covariance estimate:

$$\vec{x}_{t|t-1} = \Phi_t \vec{x}_{t-1|t-1} \tag{7}$$

$$P_{t|t-1} = \Phi_t P_{t-1|t-1} \Phi_t' + Q_t \tag{8}$$

After linearization of Equation (6) at predicted state $\vec{x}_{t|t-1}$,

$$\vec{e}_{t|t-1} \approx H_t(\vec{x}_t - \vec{x}_{t|t-1}) + \vec{v}_t \tag{9}$$

where $H_t = \left. \frac{\partial \vec{y}}{\partial \vec{x}} \right|_{\vec{x}_{t|t-1}}$, $\vec{e}_{t|t-1} = \vec{y}_t - h(\vec{x}_{t|t-1})$. $\vec{e}_{t|t-1}$ is called innovations or measurement residuals, then the Kalman filter is obtained:

$$K_t = P_{t|t-1} H_t' \left(H_t P_{t|t-1} H_t' + R_t \right)^{-1} \tag{10}$$

$$\vec{x}_{t|t} = \vec{x}_{t|t-1} + K_t \vec{e}_{t|t-1} \tag{11}$$

$$P_{t|t} = (I - K_t H_t) P_{t|t-1} \tag{12}$$

where $\vec{x}_{t|t}$ and $P_{t|t}$ are the updated Kalman estimate and the updated covariance estimate.

2.4. Kalman Smoothing

The Kalman smoother estimator could be obtained [25]:

$$\vec{x}_{t-1|N} = \vec{x}_{t-1|t-1} + J_{t-1}\{\vec{x}_{t|N} - \vec{x}_{t|t-1}\} \tag{13}$$

$$P_{t-1|N} = P_{t-1|t-1} + J_{t-1}(P_{t|N} - P_{t|t-1})J'_{t-1} \tag{14}$$

where $J_{t-1} = P_{t-1|t-1}\phi_t'[P_{t|t-1}]^{-1}$, $1 \leq t \leq N$, N is the number of epochs.

Kalman lag-one covariance holds with the starting condition

$$P_{N,N-1|N} = (I - K_N H_N)\Phi_N P_{N-1|N-1} \tag{15}$$

for $t = N, N-1, \ldots, 2$

$$P_{t-1,t-2|N} = P_{t-1|t-1}J'_{t-2} + J_{t-1}(P_{t,t-1|N} - \Phi_t P_{t-1|t-1})J'_{t-2} \tag{16}$$

3. EM-PPP

The EM-algorithm is based on the innovation of the likelihood function to compute maximum likelihood estimation [25,26]. The likelihood function describes the probability of the observations, given a set of parameters. The parameters are found such that they maximize the likelihood function. The derivative of the likelihood function or log-likelihood is not always tractable. Therefore, iterative methods like Expectation-Maximization algorithms are very effective to find numerical solutions for the parameter estimates.

Denoting $\Theta = \{\vec{\mu}_0, P_0, Q_t, R_t | t = 1, \ldots, N\}$, $X = \{\vec{x}_0, \vec{x}_1, \ldots, \vec{x}_N\}$, $Y = \{\vec{y}_1, \ldots, \vec{y}_N\}$. Y is thought of as incomplete data, and $\{X, Y\}$ as complete data. Specifically for PPP, the log likelihood of the parameters of the state space model is approximately derived (ignoring constant):

$$\begin{aligned} 2\log L_{X,Y}(\Theta) = &-\log|\Sigma_0| - (\vec{x}_0 - \vec{\mu}_0)'\Sigma_0^{-1}(\vec{x}_0 - \vec{\mu}_0) \\ &- \sum_{t=1}^{N}\log|Q_t| - \sum_{t=1}^{N}(\vec{x}_t - \Phi_t\vec{x}_t)'Q_t^{-1}(\vec{x}_t - \Phi_t\vec{x}_t) \\ &- \sum_{t=1}^{N}\log|R_t| - \sum_{t=1}^{N}(\vec{y}_t - h(\vec{x}_{t|N}))'R_t^{-1}(\vec{y}_t - h(\vec{x}_{t|N})) \end{aligned} \tag{17}$$

Since the hidden states $\vec{x}_t$ are unknown, only the expected value of the log likelihood conditioned on Y is accessible, as a result, the observation equation is expanded at smoother point $\vec{x}_{t|N}$.

The Expectation (E-step) of EM algorithm for PPP requires computing the expected log-likelihood at the *jth* iteration:

$$\Omega(\Theta|\Theta^{(j-1)}) = E\{2\log L_{X,Y}(\Theta)|Y, \Theta^{(j-1)}\} \tag{18}$$

then the parameters are recalculated at the Maximization step (M-step):

$$\Theta^{(j)} = \underset{\Theta}{\operatorname{argmax}}\ \Omega(\Theta|\Theta^{(j-1)}) \tag{19}$$

The two steps are repeated until the $\Theta^{(j)}$ converges.

The EM-PPP is terminated when the following convergence criterion is reached:

$$R\text{-log} = \left|\frac{\ell^{(j)} - \ell^{(j-1)}}{\ell^{(j)}}\right| < \varepsilon \text{ or } j \geq maximum\ number\ of\ iterations \tag{20}$$

where ε is a small predefined amount and $\ell^{(j)}$ is equal to

$$\ell^{(j)} = \sum_{t=1}^{N} \log|H_t P_{t|t-1} H_t' + R_t| + \sum_{t=1}^{N} (\vec{e}_{t|t-1})' (H_t P_{t|t-1} H_t' + R_t)^{-1} (\vec{e}_{t|t-1}) \tag{21}$$

A flowchart of our EM-PPP procedure is shown in Figure 1. In the initialization step, GNSS data preprocessing is performed including data integrity checking, cycle slips and outliers detection, phase center offset (PCO) and phase center variations (PCV) correction, synchronization of receiver clock using only code range measurements and initialization of parameters Θ. It also initializes the hidden state $\vec{x}_0$, sharing of the same processing noise Q_t across the time step $t = 1, \ldots, N$.

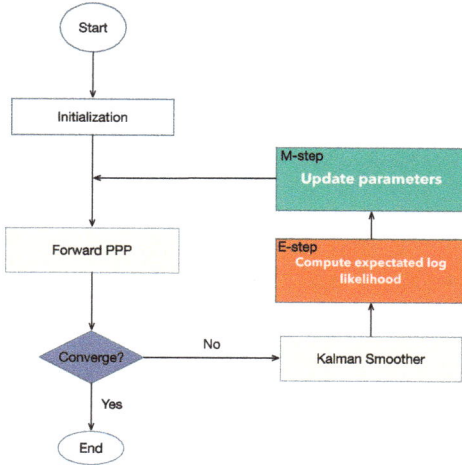

Figure 1. Flowchart of the Expectation-Maximization (EM)-Precise Point Positioning (PPP) loop.

In the next step, the extended Kalman filter Equations (10)–(12) are implemented to compute a series of hidden states and their covariance. If the convergence condition Equation (20) is not satisfied, then Kalman smoothing is used to calculate smoothed state estimates and involved covariance matrix Equations (13)–(16), which is prepared for the E-step: calculation of the expected log likelihood function Equation (22). All parameters of Θ are updated during the M-step and prepared for the next iteration Equation (27).

3.1. E-Step

Taking the expectation upon Equation (17) over conditional distribution of the latent given observed data, we find immediately:

$$\begin{aligned}
\Omega(\Theta|\Theta^{(j-1)}) &= E\{-2\ln L_{X,Y}(\Theta)|Y\} \\
&= \ln|\Sigma_0| + tr[\Sigma_0^{-1} E\{(\vec{x}_0 - \vec{\mu}_0)(\vec{x}_0 - \vec{\mu}_0)'|Y\}] + \sum_{t=1}^{N} \ln|Q_t| \\
&+ \sum_{t=1}^{N} tr[Q_t^{-1} E\{(\vec{x}_t - \Phi_t \vec{x}_{t-1})(\vec{x}_t - \Phi_t \vec{x}_{t-1})'|Y\}] + \sum_{t=1}^{N} \ln|R_t| \\
&+ \sum_{t=1}^{N} tr[R_t^{-1} E\{(\vec{y}_t - h(\vec{x}_t))(\vec{y}_t - h(\vec{x}_t))'|Y\}]
\end{aligned} \tag{22}$$

where

$$E\{(\vec{x}_0 - \vec{\mu}_0)(\vec{x}_0 - \vec{\mu}_0)'|Y\} = P_{0|N} + (\vec{x}_{0|N} - \vec{\mu}_0)(\vec{x}_{0|N} - \vec{\mu}_0)' \tag{23}$$

$$E\{(\vec{x}_t - \Phi_t \vec{x}_{t-1})(\vec{x}_t - \Phi_t \vec{x}_{t-1})'|Y\}$$
$$= P_{t|N} + \vec{x}_{t|N}\vec{x}'_{t|N} + \Phi_t(P_{t-1|N} + \vec{x}_{t-1|N}\vec{x}'_{t-1|N})\Phi'_t \qquad (24)$$
$$-\Phi_t(P_{t,t-1|N} + \vec{x}_{t|N}\vec{x}'_{t-1|N})' - (P_{t,t-1|N} + \vec{x}_{t|N}\vec{x}'_{t-1|N})\Phi'_t$$

Using Taylor series expression

$$\vec{y}_t \approx h(\vec{x}_{t|N}) + H_{t|N}(\vec{x}_t - \vec{x}_{t|N}) + \vec{v}_t, \qquad (25)$$

where $H_{t|N} = \left.\frac{\partial \vec{y}}{\partial \vec{x}}\right|_{\vec{x}_{t|N}}$, and let $\vec{e}_{t|N} = \vec{y}_t - h(\vec{x}_{t|N})$, $\Delta \vec{x}_{t|N} = \vec{x}_t - \vec{x}_{t|N}$, we get

$$E\{(\vec{y}_t - h(\vec{x}_t))(\vec{y}_t - h(\vec{x}_t))'|Y\} \approx \vec{e}_{t|N}\vec{e}'_{t|N} + H_{t|N}P_{t|N}H'_{t|N} \qquad (26)$$

3.2. M-Step

Similar to complete-data weighted maximum likelihood estimation, from the first differential of $\Omega(\Theta|\Theta^{(j-1)})$, the maximum likelihood estimators are updated as follows:

$$\begin{aligned} \vec{\mu}_0 &= \vec{x}_{0|N} \\ \Sigma_0 &= P_{0|N} \\ Q_t &= E\{(\vec{x}_t - \Phi_t \vec{x}_{t-1})(\vec{x}_t - \Phi_t \vec{x}_{t-1})'|Y\} \\ Q_t &= E\{(\vec{x}_t - \Phi_t \vec{x}_{t-1})(\vec{x}_t - \Phi_t \vec{x}_{t-1})'|Y\} \end{aligned} \qquad (27)$$

For simplicity, the initial covariance P_0, and the measurement covariance R_t are assumed to be a diagonal matrix:

$$P_0 = diag(q_{01}, q_{02}, \ldots, q_{0k_0})$$
$$R_t = diag(r_{t1}, r_{t2}, \ldots, t_{tm_t})$$

where k_0 indicates the dimension of the hidden state vector at initial epoch and m_t is the dimension of the observation vector at epoch t.

4. EM Compared to MLE and LS-VCE

In literature [19], a comprehensive comparison is demonstrated between different estimation principles such as LS-VCE, best linear unbiased estimator (BLUE), best invariant quadratic unbiased estimator (BIQUE), minimum norm quadratic unbiased estimator (MINQUE) and restricted maximum likelihood estimator (REML). As shown previously in Section 3.1, the EM algorithm may be thought of as maximum likelihood estimation (MLE), but which finds the ML estimator in an iterative way. EM can be realized based on REML as well. Therefore, an additional comparison between EM variance estimation and LS-VCE and MLE is adequate.

To make theoretical analysis easy and consistent, in the following we first introduce how to covert Kalman filter to least-squares. Then we directly give different (co)variance estimators according to their distribution assumptions and the reason why the EM-algorithm is preferable in our solution.

4.1. From Kalman Filter to Least Squares

The linear (extended) Kalman equation can be transformed into the least-squares function model, which allows the following EM algorithm to be compared with LS-VCE on the same function model and makes the theoretical analysis easy and convenient. To do so, the state Equation (1) and the observation Equation (6) are expanded at a priori value and organized as the function model:

$$\vec{y} = A\vec{x} + \vec{w}$$
$$E\{\vec{w}\} = 0,\ D\{\vec{w}\} = E\{\vec{w}\vec{w}'\} = Q = Q_0 + \sum_{k=1}^{p} \sigma_k Q_k \tag{28}$$

with

$$\vec{y} = [\ 0\ \ d\vec{y}_1{}'\ \ 0\ \cdots\ 0\ \ d\vec{y}_N{}'\ ]'$$

$$A = \begin{bmatrix} \Phi_1 & -1 & 0 & \cdots & 0 & 0 \\ 0 & H_1 & 0 & \cdots & 0 & 0 \\ 0 & \Phi_2 & -1 & \cdots & 0 & 0 \\ \vdots & \vdots & \vdots & \ddots & \vdots & \vdots \\ 0 & 0 & 0 & \cdots & \Phi_N & -1 \\ 0 & 0 & 0 & \cdots & 0 & H_N \end{bmatrix}$$

$$\vec{x} = [\ d\vec{x}_0{}'\ \ d\vec{x}_1{}'\ \ d\vec{x}_2{}'\ \cdots\ d\vec{x}_{N-1}{}'\ \ d\vec{x}_N{}'\ ]'$$

where the a priori $\vec{x}_i^0$, $(i = 0, \ldots, N)$ value is subtracted from the original state vector $\vec{x}_i$, leading to $d\vec{x}_i = \vec{x}_i - \vec{x}_i^0$ and $d\vec{y}_i = \vec{y}_i - \vec{y}_i^0$, N is the number of epochs. The $m \times n$ matrix A is full column rank. The cofactor matrices Q_k, are assumed to be known and their weighted sum $Q_0 + \sum_{k=1}^{p} \sigma_k Q_k$ is assumed to be positive definite and Q_k, $(k = 1, \ldots p)$ are linearly independent, and the (co)variance components σ_k are unknown parameters. Matrix Q_0 is the known part of the variance matrix [19].

4.2. Least-Squares EM

Similar to Section 3, we can calculate the non-constant part of the full log-likelihood function and then conditional expectation on the observation $\vec{y}$ and $Q^{(j)}$, given the data $\vec{y}$ and the jth iteration estimates of (co)variance components $\sigma_k^{(j)}$ or $Q^{(j)}$:

$$\begin{aligned}\Omega(Q|\vec{y}, Q^{(j)}) &= E\{L|\vec{y}, Q^{(j)}\} = \log Q + E\{tr((\vec{y} - A\vec{x})'Q^{-1}(\vec{y} - A\vec{x}))|\vec{y}\} \\ &= \log Q + tr(Q^{-1}E\{(\vec{y} - A\vec{x})(\vec{y} - A\vec{x})'|\vec{y}\})\end{aligned} \tag{29}$$

where

$$E\{(\vec{y} - A\vec{x})(\vec{y} - A\vec{x})'|\vec{y}\} = (\vec{y} - A\hat{x})(\vec{y} - A\hat{x})' + AQ_{\hat{x}}^{(j)}A'Z_{ML}^{(j)}$$
$$\hat{x} = (A'(Q^{(j)})^{-1}A)^{-1}A'(Q^{(j)})^{-1}\vec{y} \tag{30}$$

M-step:

Maximizing the likelihood of the completed data based on Equation (29), the new estimates $\sigma_k^{(j+1)}$ are calculated as

$$\begin{bmatrix} \sigma_1^{(j+1)} \\ \vdots \\ \sigma_p^{(j+1)} \end{bmatrix} = (A'_{vEM}W_{vEM}A_{vEM})^{-1}A'_{vEM}W_{vEM}\ vec\left(Z_{EM}^{(j)} - Q_0\right) \tag{31}$$

where $A_{vEM} = \begin{bmatrix} vec\left(Q_1^{(j)}\right) & \cdots & vec\left(Q_p^{(j)}\right) \end{bmatrix}$, $W_{vEM} = (Q^{(j)})^{-1} \otimes (Q^{(j)})^{-1}$, $\otimes$ is the Kronecker product and vec is *vec*-operator.

Equations (29) and (31) are the EM algorithm for ML estimation. If convergence is reached, set $\sigma_k = \sigma_k^{(j)}$, otherwise increase j by one and return to E-step.

4.3. MLE

Once the general structure of probability density function is known, MLE can be simply realized and therefore used widely. If a multivariate normal distribution is given, the (co)variance components takes form:

$$\begin{bmatrix} \sigma_1 \\ \vdots \\ \sigma_p \end{bmatrix} = (A'_{vMLE} W_{vMLE} A_{vMLE})^{-1} A'_{vMLE} W_{vMLE} \, vec(Z_{MLE} - Q_0) \tag{32}$$

where $A_{vMLE} = \begin{bmatrix} vecQ_1 & \cdots & vecQ_p \end{bmatrix}$, $W_{vMLE} = Q^{-1} \otimes Q^{-1}$.

From Equations (31) and (32), we know that least-squares EM and MLE estimators share the same design matrix and weight matrix. Their difference is mainly caused by the pseudo observation $vec\left(Z_{EM}^{(j)} - Q_0\right)$ and $vec(Z_{MLE} - Q_0)$. $Z_{EM}^{(j)}$ for EM-algorithm includes the effects of both observation post-residuals and the accuracy of the estimates $\hat{x}$. In contrast, Z_{MLE} for MLE consider only the observation post-residuals. Therefore, MLE estimator is probably over-optimistic to EM.

However, if the REML principle is used to derive the EM-algorithm, the effect of $\hat{x}$ is implicitly removed. Then, the EM-algorithm based on REML will be equivalent to the REML estimator.

4.4. LS-VCE

Another important problem is that the EM-algorithm and MLE do not take the loss of degrees of freedom from the estimation of $\vec{x}$ into account. Borrowing the idea of REML, LS-VCE overcomes this problem based on $(n-p)$ independently error contrasts. Specifically, let

$$\hat{t} = B'\vec{y}, Q_{\hat{t}} = B'QB \tag{33}$$

where $\vec{t}$ is misclosure vector, B is $m \times (m-n)$ matrix satisfying $B'A = 0$, $rank(B) = m - n = b$. Then LS-VCE estimator is obtained:

$$\begin{bmatrix} \sigma_1 \\ \vdots \\ \sigma_p \end{bmatrix} = (A'_{vLS-VCE} W_{vLS-VCE} A_{vLS-VCE})^{-1} A'_{vLS-VCE} W_{vLS-VCE} \, vec(Z_{LS-VCE} - Q_0) \tag{34}$$

with $A_{vLS-VCE} = \begin{bmatrix} vec(B'Q_1B) & \cdots & vec(B'Q_pB) \end{bmatrix}$, $Z_{LS-VCE} = \hat{t}\hat{t}'$, $W_{vLS-VCE}$ is a user-defined weight matrix. If $W_{vLS-VCE}$ is set to $Q_{\hat{t}} \otimes Q_{\hat{t}}$, LS-VCE is the same as EM-algorithm based on REML.

LS-VCE is derived purely based on the least-squares method and we do not make any assumption on a probability density function (PDF). In contrast, EM-algorithm, MLE and REML are built upon a certain distribution, which explains why when applying LS-VCE, it is necessary for users to set weight matrix on their own.

4.5. Preference for Recursive EM

As discussed previously, the EM-algorithm can be implemented as either recursive form or batch form like MLE and LS-VCE. In our solution, we prefer the EM-algorithm based Kalman filter to other methods.

Recursive EM discriminates between the process noise and the observation noise. For GNSS, the process noise is usually different from observation noise. The process noise is directly connected to the geophysical phenomenon, which has not only linear but also non-linear variations, and suffers both time and spatial correlations [27–29]. As a result, it is relatively more difficult to estimate the process noise than the observation covariance matrix. Other batch methods mix the two types of stochastic processes with different behavior, which will bring us extra trouble.

Recursive EM can process data in not only post mode, but also in real-time mode. In both modes, data is processed epoch by epoch, allowing us to dynamically adjust the weight matrix, monitoring time-varying behavior and detecting abrupt motion.

5. Validation

5.1. Static PPP Scheme

A total of 14 IGS multi-GNSS stations are selected to assess the performance of EM-PPP (Figure 2). Those stations are evenly distributed on the Earth and track as many GNSS constellations as possible, covering not only the continent but also coastal and polar regions.

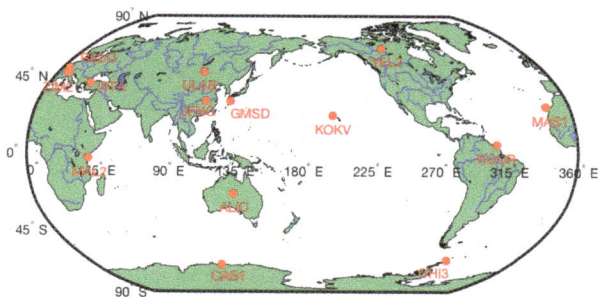

Figure 2. Distribution of selected IGS multi-GNSS stations.

Daily GNSS measurements from those IGS stations observed during DOY 119 in 2017 are used in this study. The true coordinate benchmarks are from IGS weekly solutions. The final GFZ Beidou multi-GNSS (GBM) products of the satellite orbits and clocks are applied (ftp://cddis.gsfc.nasa.gov/pub/gps/products/mgex/). The used precise orbit correction has a sampling interval of 15 min while the precise clock has a sampling interval of 30 s, both generated at GFZ. For GPS, we aligned C1 to P1 using CODE differential code bias (DCB) product (ftp://ftp.aiub.unibe.ch/CODE/2017/). The frequency bands we used are L1 and L2 for GPS, G1 and G2 for Glonass, E1 and E5a for Galileo, B1 and B2 for BeiDou [9]. Receiver and satellite PCO and PCV were corrected using igs14.atx. solid Earth tides, pole tide and ocean tides are removed according to IERS Conventions 2010. For troposphere delay estimation, the zenith dry component of tropospheric delays was corrected with the a priori Niell model [24]. The zenith wet delay (ZWD) is estimated as an unknown parameter. Then, 24-h observation data sets with a sampling interval of 30 s were processed for all stations. The elevation cut-off was set to 6 degrees.

The initial guess of receiver coordinates is intentionally deviated by 100 m from IGS station solution. A priori standard deviation (std) of PC is set to 3 m, a priori std of LC to 0.03 m for pseudo-range combination (PC) and carrier phase combination (LC) combination, respectively.

The starting values for Q_t are shown in Table 1. Random walk noise process with a spectral density equal to 1.0 mm/$\sqrt{30}$ s is imposed on coordinates, which means a 1.0 mm disturbance very 30 s for IGS station in North, East and Up. It is not true in reality of course, but useful for test purposes. The receiver clock offset is supposed to be white noise. Zenith wet delay (ZWD) and inter-system bias (ISB) are modeled as random walk noise. Ambiguities can be considered as constant or random walk noise with very small spectral density.

Table 1. Initial guess Q_t for static PPP: ZWD is the zenith wet delay of the troposphere, ISB is the system time difference with respect to GPS time.

	X	Y	Z	cdt_r	ZWD	ISB	Ambiguities
Q_t	$\frac{1.0 \text{ mm}}{30 \text{ s}}$	$\frac{1.0 \text{ mm}}{30 \text{ s}}$	$\frac{1.0 \text{ mm}}{30 \text{ s}}$	9.0×10^{10} m	$\frac{1.0 \text{ cm}}{\sqrt{h}}$	$\frac{3.0 \times 10^{-6} \text{ m}}{\sqrt{s}}$	0.0

If the maximum number of iteration is reached and EM-PPP does not converge, smaller spectral density should be assigned for X, Y and Z for the next cycle.

5.2. Kinematic PPP Scheme

Another kinematic dataset was used to further validate the performance of the EM-PPP. The data was collected at Wuhan, China, in November 14, 2013. The sampling interval is 1 Hz and the observed time span is about one hour. The final CODE precise satellite orbit and 5 s clock products are used to estimate the 1 Hz GPS displacements. The ambiguity-fixed double differenced real-time kinematic (RTK) solutions are adopted as the reference to assess the performance of kinematic EM-PPP solution.

The initial acceleration variance is assumed to be $\frac{10 \text{ m}^2}{s^3}$ for position and velocity states (Table 2), which can be used to calculate the process noise matrix for position and velocity. The initial Q_t for receiver clock is modeled as white noise and estimated epoch-wisely. ZWD and ambiguities are also modeled as random walk processes with initial spectral density 0.01 cm/$\sqrt{s}$ and 0.0 m/$\sqrt{s}$, respectively.

Table 2. Initial guess Q_t for kinematic PPP: ZWD is zenith wet delay of troposphere.

	q_x	q_y	q_z	cdt_r	ZWD	Ambiguities
Q_t	$\frac{10 \text{ m}^2}{s^3}$	$\frac{10 \text{ m}^2}{s^3}$	$\frac{10 \text{ m}^2}{s^3}$	$(100 \text{ m})^2$	0.01 cm/$\sqrt{s}$	0.0

6. Results and Discussion

6.1. Static EM-PPP Solution

It is found that EM-PPP usually converges after the iteration counter reaches 50. The positioning errors are shown in Table 3, including PPP results at 1st iteration with the biased stochastic model, and the results after 50 iterations of calibration to assess EM-PPP performance. EM-PPP convergence in our research means that the square root of 3D positioning errors of the last 20 consecutive epochs is less than 10 cm.

Table 3 indicates that when the biased Q_t and R_t are fed in the beginning, PPP 3D errors are up to decimeters for a few stations. Horizontal errors are often greater than vertical errors, which is not consistent with the property of GNSS, because of inappropriate process noise matrix and measurement noise covariance matrix. After 50 iterations, the position errors are reduced to within 1 cm in North, East and Up direction on average using our EM-PPP algorithm. The mean 3D error is reduced to 1.77 cm without fixing ambiguities. The overall decrease percentage on average is 66.91%, 66.16%, 71.60% in North, East and Up direction, respectively, 69.95% for 3D errors.

Table 3. Statistics of EM-PPP absolute positioning errors (cm) with respect to IGS stations solution. All sites except YEL2 are processed using multiple GNSS observations. YEL2 is processed using only GPS data to verify the algorithm for single GNSS constellation. The ambiguities are not fixed.

No.	Site	1st Iteration				50th Iteration			
		North	East	Up	3D	North	East	Up	3D
1	ALIC	4.32	12.17	4.38	13.64	0.89	1.47	0.24	1.74
2	CAS1	2.49	2.23	2.97	4.48	1.58	0.80	0.91	2.00
3	GMSD	0.29	0.13	2.38	2.41	0.22	0.59	1.22	1.38
4	ISTA	2.11	1.44	1.02	2.75	2.11	0.07	0.95	2.33
5	JFNG	3.32	3.65	5.48	7.39	0.75	0.52	0.64	1.12
6	KOKV	0.88	4.97	3.43	6.11	0.54	2.13	3.20	3.89
7	KOUR	0.15	2.38	2.27	3.30	1.06	0.76	0.39	1.37
8	MAL2	1.21	2.99	5.67	6.53	1.23	1.50	0.98	2.18
9	MAS1	3.38	1.15	1.67	3.94	1.36	1.22	1.00	2.09
10	OHI3	16.47	2.94	4.83	17.42	1.13	0.44	0.55	1.34
11	REDU	0.76	0.39	5.19	5.25	0.56	0.59	0.40	0.91
12	ULAB	1.47	0.04	0.19	1.48	0.71	0.60	0.30	0.98
13	YEL2	0.61	2.17	2.72	3.54	0.34	0.93	0.12	1.00
14	ZIM2	0.15	0.49	4.15	4.19	0.03	1.02	2.25	2.47
	average	2.69	2.66	3.31	5.89	0.89	0.90	0.94	1.77

To see what happened to the process noise Q_t and the observation covariance matrix R_t before and after calibration, an example of JFNG station located in China is illustrated.

The residuals for PC and the corresponding formal errors are shown in Figure 3. The residuals for LC and the corresponding formal errors against satellite elevation angles after calibration are shown in Figure 4. To be clear, PC and LC residuals for BeiDou are plotted separately from those for GPS, Glonass and Galileo.

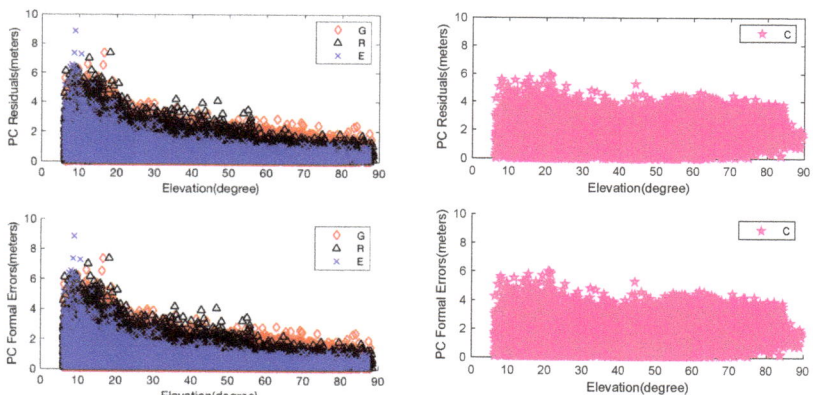

Figure 3. EM-PPP absolute pseudo-range combination (PC) residuals and formal errors of JFNG station at 50th iteration: Left two pictures are the PC residuals and their formal errors (square root of observation matrix R_t) of GPS, Glonass and Galileo, respectively. Similarly, the right two pictures show the PC residuals and formal errors of BeiDou.

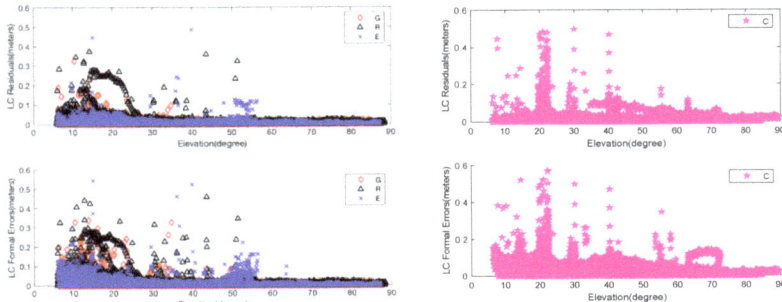

Figure 4. EM-PPP carrier phase combination (LC) residuals and formal errors of JFNG station at 50th iteration: Left two pictures are the LC residuals and formal errors (square root of the observation covariance matrix) for GPS, Glonass and Galileo, respectively. Similarly, the right two pictures show the LC residuals and formal errors for BeiDou.

The results allow us to examine the relationship between R_t and the residuals. It is clearly shown that formal errors of R_t are changed from 3 m to vary about between 0 and 10 m for PC, and from 0.03 m to vary about between 0 and 0.6 m for LC. The big outliers of LC are due to the non-convergence at the initial epoch and EM-PPP does not fully converge at some epochs. Although R_t variations, similar to the residuals-varying pattern, are dependent on the satellite elevation angle, LC formal error of BeiDou R_t at the elevation of about 20 degrees is greater than the lower degree formal error (refer to the bottom right picture in Figure 4). Clearly, it is not advisable to choose observation weight only according to the satellite elevation angle. Another example is station ISTA, where Glonass R_t for PC at satellite elevation higher than 50 degrees is almost as great as lower degree errors (not plotted here). In addition, random or systematic outliers are downweighed accordingly for both PC and LC.

It can also be observed that BeiDou PC residuals peak are almost 6 m, worse than GPS, Glonass and Galileo at JFNG. In fact, the biggest error source comes from GEO C05 and IGSO C06, C07 and C08, probably due to their poor orbit accuracy and clock offset, because the residuals of those satellites show less independence of satellite elevation angle. R_t is different among GPS, Glonass and Galileo.

Thus R_t not only reflects the accuracy of measurement type itself, elevation-dependent, characteristics of GNSS constellation, but also the quality of GNSS satellite orbit and clock products, and should be adjusted dynamically. Consequently, EM-PPP is effective to calibrate R_t automatically and suppress outliers.

In general, Q_t can be adjusted to its correct value, which is zero in our case, after less than ten iterations, as shown in Figure 5. The initial square root of Q_t in North, East and Up directions are 0.001 m/ $\sqrt{30}$ s. Q_t becomes zero at the sixth iteration in all three components and position solution at the last epoch varies little.

It has to be pointed out that EM-PPP suffers local extrema problems like other alternative methods. It is not an algorithm to locate the global maxima, therefore EM-PPP is sensitive to initial guesses. To escape from this local extrema trap, several sets of initialization schemes can be used, and select the best one selected as the final result.

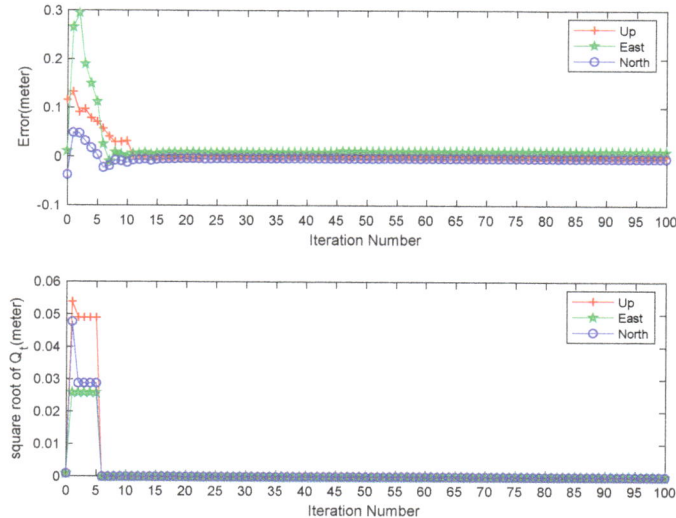

Figure 5. Effects of EM-PPP iteration number on the position errors (top) and variations of the square root of Q_t at last epoch on station JFNG.

6.2. Kinematic EM-PPP Solution

Figure 6 is the tracking route recovered by the EM-PPP at the 500th iteration. The relative log-likelihood versus iterations is plotted in Figure 7, where the results at the first iteration correspond to the solution to the traditional PPP (TPPP). The relative log likelihood decreases gradually from 1.0 to 8.665×10^{-7} and forms a concave curve. The less the relative log-likelihood, the less perturbative the EM-PPP solution becomes with respect to the RTK solution. As the iterative number grows to about 150, the EM-PPP solution converges nearly completely.

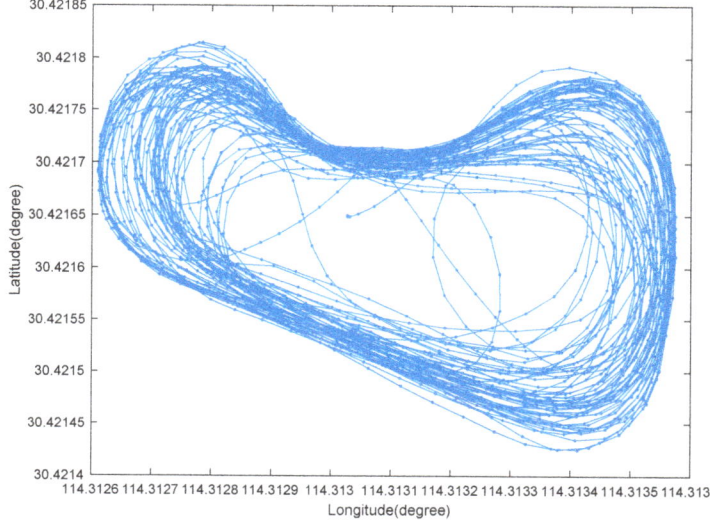

Figure 6. Driving route of moving vehicle in Wuhan, China on 14 November 2013.

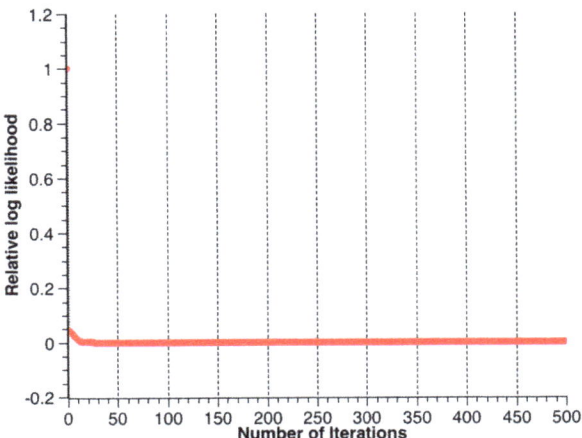

Figure 7. Relative log-likelihood of two consecutive iterations.

Illustrated in Figure 8 are the positioning errors of the traditional PPP solution and EM-PPP solution at the 500th iteration with respect to the reference coordinates in the Up, East and North directions. It is observed that EM-PPP solutions are much more stable in all directions when compared to TPPP solutions. The TPPP solution in the East changes wobbly in comparison to the North and Up due to greater acceleration in the East (Figure 9), which leads to a larger bias in the East.

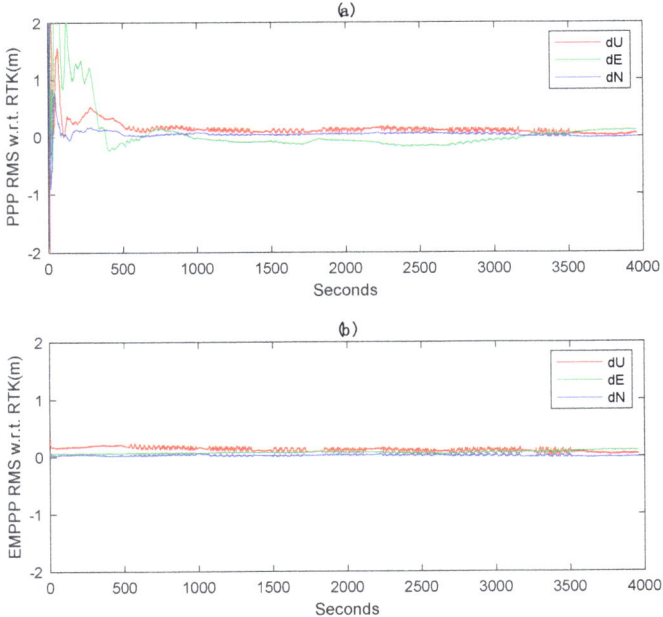

Figure 8. GPS-only kinematic positioning errors with respect to real-time kinematic (RTK): (**a**) traditional PPP errors, (**b**) EM-PPP positioning errors at the 500th iteration. Traditional PPP and EM-PPP share the same initial conditions.

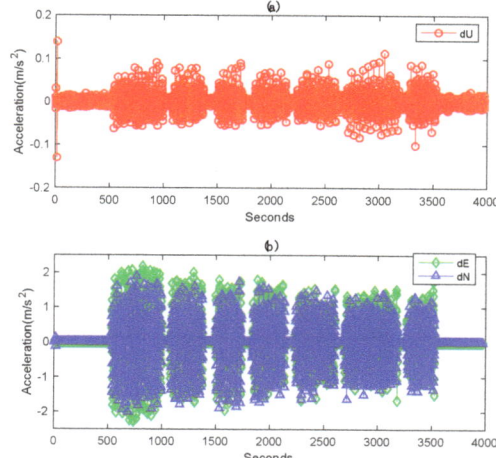

Figure 9. Acceleration time series derived from second-order differencing RTK position time series in Up, East and North: (**a**) acceleration in Up. (**b**) Accelerations in the East and North.

In fact, the position and velocity (PV) model of our kinematic state equation is assumed to be a constant velocity model, which means the acceleration is zero. However, the realistic acceleration of the vehicle is no zero, which results in systematic bias and unstable solution.

Figures 10 and 11 describe the EM-PPP RMSs and STDs with respect to RTK against the number of iterations, respectively. Obviously, after about 40 iterations, RMSs in Up and North are decreasing and the solutions are improved in those two directions. In contrast to Up and North direction, in the East direction it presents a decrease from 9.7 cm falling to 7.6 cm and then increases up to 8.5 cm for the RMS. However, the combined effects of all three directions get decreasingly 3D RMS, proving that EM-PPP does improve the positioning accuracy in the kinematic mode in our case, and apparently converges with increasing iterations, consistent with the EM theory, though there is a little disturbance because of the existing system bias and outliers of pseudo-range. It can be imagined that a better result can be expected if acceleration observations are also observed.

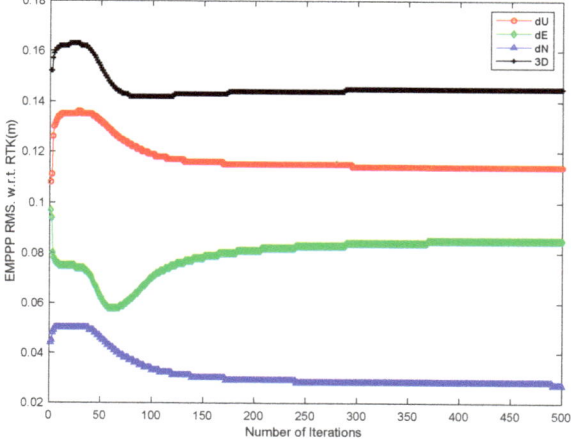

Figure 10. EM-PPP RMSs w.r.t. RTK solution in Up, East and North (unit: m).

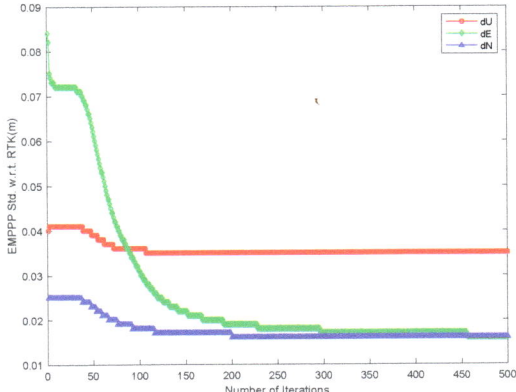

Figure 11. EM-PPP STDs w.r.t. RTK solution in Up, East and North (unit: m).

As for STDs with respect to RTK, they are consistently increasing in all three directions when EM-PPP continues its iteration. The STDs decrease from 4.0 cm, 8.4 cm and 2.5 cm to 3.5 cm, 1.6 cm and 1.6 cm in Up, East and North components, respectively. In other words, the STDs are improved by 12.5%, 80.9% and 36.0% in Up, East and North, accordingly.

Given Figure 12 is the estimates of geodetic coordinates B, L and H, and their estimates of the square root of Q_t after the 500th iteration. It is noted that the coordinates simultaneously stay stable or alternatively change sharply in all three directions, telling that the moving patterns of the vehicle switch between static and kinematic status from time to time. Theoretically, the process Q_t should change between zero and positive values. Obviously, the estimates of the square root Q_t, displayed in Figure 12b–f agree well with variations of coordinates. Time-varying Q_t is identified, which takes the concrete dynamic mode into consideration.

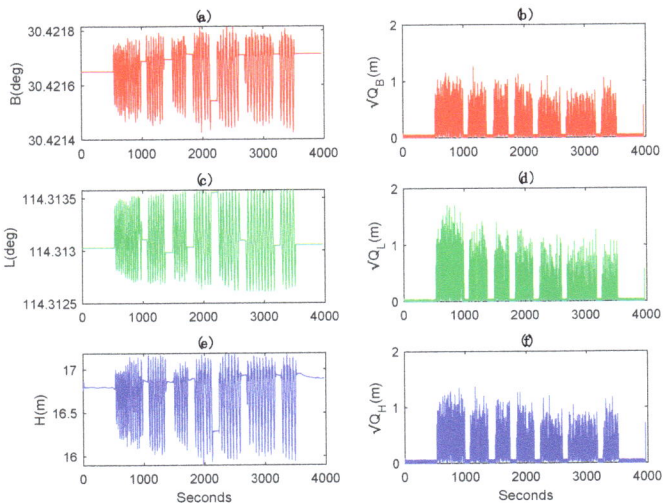

Figure 12. EM-PPP estimates of the Geodetic coordinates B, L, H and their correspondingly calibrated Q_t after the 500th iteration: (**a**) Latitude estimates B (degree). (**b**) The square root of Q_t of B (m). (**c**) Longitude estimates L (degree). (**d**) The square root of Q_t of L (m). (**e**) Height estimates H (m). (**f**) The square root of Q_t of H (m).

If much smaller initial acceleration variances are given, for example $\frac{0.01 \, m^2}{s^3}$ rather than $\frac{10 \, m^2}{s^3}$ for position and velocity states, it can be shown that our algorithm can still recover the process noise matrix as well and makes no difference. As a result, the user can choose values for Q_t randomly to a large extent.

7. Conclusions

A machine learning algorithm, EM algorithm, is adapted particularly to the extended Kalman filter to calibrate the process noise matrix and observation covariance noise matrix for PPP. The main advantage of EM-PPP is in fact that it is straightforward, simple, (locally) optimal and able to estimate large amounts of parameters and thus competent in calibrating the time-varying process noise and observation covariance for PPP state-space model, though its execution is time-consuming. The basic framework of EM-PPP is not limited to multi-PPP and can be applied to other fields of geodesy.

The whole procedure of EM-PPP is comprised of three parts: initialization, feedforward and backpropagation. In the beginning, the GNSS preprocess is performed to check the availability of the required data and mainly recognize cycle slips. Next, the whole process iterates between the estimation of hidden state and expectation and maximization.

The EM-algorithm is then compared with MLE and LS-VCE methods. We choose the recursive algorithm because it is superior to separate the process noise and observation variance, and to monitor time-varying behavior.

The approach was verified by selecting a global distribution of 14 IGS multi-GNSS station without fixing ambiguities. Based on the presented results, it concluded that EM-PPP is well suited for dynamically determining the time-varying process noise and observation noise. The calibrated observation variance matches the observation residuals from low satellite elevation angle to high satellite elevation angle. It resists orbit and clock errors and outliers through downweighing abnormal observations at different epochs, which is an alternative reasonable solution in contrast to the popular way that assigns weight according to the satellite elevation angle. People do not need to worry about separating observations into different categories carefully based on different GNSS constellations to estimate variance components like variance component estimation (VCE).

The spectral density of the assumed kinematic IGS station with 1 mm disturbance every 30 s in North, East and Up direction was estimated to be zero, implying that stations are static, which is consistent with reality.

An additional kinematic test was also implemented and reasonable values of Q_t are found when biased initial Q_t guess was given. The position errors are reduced in Up, East and North direction, respectively, w.r.t. RTK. In particular the STDs with respect to RTK are improved by 12.5%, 80.9% and 36.0% in Up, East and North, further showing that EM-PPP is also beneficial to kinematic PPP.

It has been confirmed that the EM-PPP is competitive for the calibration of the PPP stochastic model dynamically. The main drawback of this approach is that it converges slowly due to its first-order convergence. In the future, online EM-PPP may be derived to process GNSS data in real-time to overcome this problem if a large number of observations are available.

Author Contributions: X.Z. conceived and designed the algorithm; P.L. provided the kinematic GNSS data and helped validate the algorithm; M.G. supervised the whole procedure, and continuously discussed and analyzed the results and gave constructive suggestions; R.T. and X.L. participated in the experimental investigation; H.S. helped edit and revise the paper. All authors have read and agreed to the published version of the manuscript.

Funding: This research is funded by the China Scholarship Council (CSC).

Acknowledgments: X.Z. is financially supported by the China Scholarship Council (CSC) for his study at the German Research Centre for Geosciences (GFZ). We thank the IGS and GFZ for providing GNSS observations, DCBs, precise orbit, and clock products. Our research is also partly supported by the Chinese Academy of Sciences (CAS) program of "Light of West China" (Grant No. 29Y607YR000103), Chinese Academy of Sciences, Russia and Ukraine and other countries of special funds for scientific and technological cooperation (Grant No. 2BY711HZ000101). This work is also partly supported by the National Natural Science Foundation of China (Grant No: 41674034, 41974032, 11903040), the Chinese Academy of Sciences (CAS) programs of "Pioneer Hundred

Talents" (Grant No: Y923YC1701), and the Chinese Academy of Sciences (CAS) program of "Western Youth Scholar" (Grant No: Y712YR4701, Y916YRa701), as well as "The Frontier Science Research Project" (Grant No: QYZDB-SSW-DQC028). We also thank the reviewers for their careful review.

Conflicts of Interest: The authors declare no conflict of interest.

References

1. Zumberge, J.F.; Heflin, M.B.; Jefferson, D.C.; Watkins, M.M.; Webb, F.H. Precise point positioning for the efficient and robust analysis of GPS data from large networks. *J. Geophys. Res. Solid Earth* **1997**, *102*, 5005–5017. [CrossRef]
2. Kouba, J.; Héroux, P. Precise point positioning using IGS orbit and clock products. *GPS Solut.* **2001**, *5*, 12–28. [CrossRef]
3. Li, P.; Zhang, X.; Ge, M.; Schuh, H. Three-frequency BDS precise point positioning ambiguity resolution based on raw observables. *J. Geod.* **2018**, *92*, 1357–1369. [CrossRef]
4. Ge, M.; Gendt, G.; Rothacher, M.A.; Shi, C.; Liu, J. Resolution of GPS carrier-phase ambiguities in precise point positioning (PPP) with daily observations. *J. Geod.* **2008**, *82*, 389–399. [CrossRef]
5. Laurichesse, D.; Mercier, F.; Berthias, J.; Broca, P.; Cerri, L. Integer ambiguity resolution on undifferenced GPS phase measurements and its application to PPP and satellite precise orbit determination. *Navigation* **2009**, *56*, 135–149. [CrossRef]
6. Collins, P.; Bisnath, S.; Lahaye, F.; Héroux, P. Undifferenced GPS ambiguity resolution using the decoupled clock model and ambiguity datum fixing. *Navigation* **2010**, *57*, 123–135. [CrossRef]
7. Seepersad, G.; Bisnath, S. Reduction of PPP convergence period through pseudorange multipath and noise mitigation. *GPS Solut.* **2015**, *19*, 369–379. [CrossRef]
8. Banville, S.; Collins, P.; Zhang, W.; Langley, R. Global and regional ionospheric corrections for faster PPP convergence. *Navigation* **2014**, *61*, 115–124. [CrossRef]
9. Li, P.; Zhang, X.; Guo, F. Ambiguity resolved precise point positioning with GPS and BeiDou. *J. Geod.* **2017**, *91*, 25–40.
10. Li, P.; Zhang, X.; Ren, X.; Zuo, X.; Pan, Y. Generating GPS satellite fractional cycle bias for ambiguity-fixed precise point positioning. *GPS Solut.* **2016**, *20*, 771–782. [CrossRef]
11. Mehra, R. On the identification of variances and adaptive Kalman filtering. *IEEE Trans. Autom. Control* **1970**, *15*, 175–184. [CrossRef]
12. Maybeck, P.S. Combined State and Parameter Estimation for On-Line Applications. Ph.D. Thesis, Massachusetts Institute of Technology, Cambridge, MA, USA, 1972.
13. Maybeck, P.S. Moving-bank multiple model adaptive estimation and control algorithms: An evaluation. *Control Dyn. Syst.* **1989**, *31*, 1–31.
14. Mohamed, A.; Schwarz, K. Adaptive Kalman filtering for INS/GPS. *J. Geod.* **1999**, *73*, 193–203. [CrossRef]
15. Odelson, B.; Lutz, A.; Rawlings, J. The autocovariance least-squares method for estimating covariances: Application to model-based control of chemical reactors. *IEEE Trans. Control Syst. Technol.* **2006**, *14*, 532–540. [CrossRef]
16. Magill, D. Optimal adaptive estimation of sampled stochastic processes. *IEEE Trans. Autom. Control* **1965**, *10*, 434–439. [CrossRef]
17. Yang, Y.; He, H.; Xu, G. Adaptively robust filtering for kinematic geodetic positioning. *J. Geod.* **2001**, *75*, 109–116. [CrossRef]
18. Yang, Y.; Gao, W. An optimal adaptive Kalman filter. *J. Geod.* **2006**, *80*, 177–183. [CrossRef]
19. Teunissen, P.J.; Amiri-Simkooei, A.R. Least-squares variance component estimation. *J. Geod.* **2008**, *82*, 65–82. [CrossRef]
20. Jamshidian, M.; Jennrich, R.I. Conjugate gradient acceleration of the EM algorithm. *J. Am. Stat. Assoc.* **1993**, *88*, 221–228.
21. Koch, K.R. Robust estimation by expectation maximization algorithm. *J. Geod.* **2013**, *87*, 107–116. [CrossRef]
22. Cai, C.; Gao, Y. Modeling and assessment of combined GPS/GLONASS precise point positioning. *GPS Solut.* **2013**, *17*, 223–236. [CrossRef]
23. Böhm, J.; Niell, A.; Tregoning, P.; Schuh, H. Global Mapping Function (GMF): A new empirical mapping function based on numerical weather model data. *Geophys. Res. Lett.* **2006**, *33*. [CrossRef]

24. Niell, A.E. Global mapping functions for the atmosphere delay at radio wavelengths. *J. Geophys. Res. Solid Earth* **1996**, *101*, 3227–3246. [CrossRef]
25. Shumway, R.H.; Stoffer, D.S. An approach to time series smoothing and forecasting using the EM algorithm. *J. Time Ser. Anal.* **1982**, *3*, 253–264. [CrossRef]
26. Dempster, A.P.; Laird, N.M.; Rubin, D.B. Maximum likelihood from incomplete data via the EM algorithm. *J. R. Stat. Soc. Ser. B (Methodol.)* **1977**, *39*, 1–22.
27. Benoist, C.; Collilieux, X.; Rebischung, P.; Altamimi, Z.; Jamet, O.; Métivier, L.; Chanard, K.; Bel, L. Accounting for spatiotemporal correlations of GNSS coordinate time series to estimate station velocities. *J. Geodyn.* **2020**, *135*, 101693. [CrossRef]
28. Bock, O.; Collilieux, X.; Guillamon, F.; Lebarbier, E.; Pascal, C. A breakpoint detection in the mean model with heterogeneous variance on fixed time intervals. *Stat. Comput.* **2020**, *30*, 195–207. [CrossRef]
29. Jongrujinan, T.; Satirapod, C. Improving the stochastic model for VRS network-based GNSS surveying. *Artif. Satell.* **2019**, *54*, 17–30. [CrossRef]

© 2020 by the authors. Licensee MDPI, Basel, Switzerland. This article is an open access article distributed under the terms and conditions of the Creative Commons Attribution (CC BY) license (http://creativecommons.org/licenses/by/4.0/).

Article

Elementary Error Model Applied to Terrestrial Laser Scanning Measurements: Study Case Arch Dam Kops

Gabriel Kerekes * and Volker Schwieger

Institute of Engineering Geodesy, University of Stuttgart, Geschwister-Scholl-Str. 24D, 70174 Stuttgart, Germany; volker.schwieger@iigs.uni-stuttgart.de
* Correspondence: gabriel.kerekes@iigs.uni-stuttgart.de

Received: 18 March 2020; Accepted: 10 April 2020; Published: 15 April 2020

Abstract: All measurements are affected by systematic and random deviations. A huge challenge is to correctly consider these effects on the results. Terrestrial laser scanners deliver point clouds that usually precede surface modeling. Therefore, stochastic information of the measured points directly influences the modeled surface quality. The elementary error model (EEM) is one method used to determine error sources impact on variances-covariance matrices (VCM). This approach assumes linear models and normal distributed deviations, despite the non-linear nature of the observations. It has been proven that in 90% of the cases, linearity can be assumed. In previous publications on the topic, EEM results were shown on simulated data sets while focusing on panorama laser scanners. Within this paper an application of the EEM is presented on a real object and a functional model is introduced for hybrid laser scanners. The focus is set on instrumental and atmospheric error sources. A different approach is used to classify the atmospheric parameters as stochastic correlating elementary errors, thus expanding the currently available EEM. Former approaches considered atmospheric parameters functional correlating elementary errors. Results highlight existing spatial correlations for varying scanner positions and different atmospheric conditions at the arch dam Kops in Austria.

Keywords: elementary error model; terrestrial laser scanning; variance-covariance matrix

1. Introduction

One of the main tasks in engineering geodesy is deformation and displacement monitoring of structures such as buildings, bridges, towers, dams, tunnels or other infrastructure works (cf. [1,2]). Independent of the measurement method, geodetic sensors are used to gather data either in a continuously manner or within different epochs. In both cases, these prerequisites are essential: a common geodetic reference system for all the epochs, knowledge about the deformation process and a stochastic model that describes the uncertainty of the measurements. Classical geodetic measurement methods like Global Navigation Satellite System (GNSS), total station, leveling, etc. have been used for decades in terrestrial point-wise monitoring and have well established and broadly accepted stochastic models [3]. Although highly reliable, point-wise acquisition methods have their limitations if objects with complex shapes like curved facades, high-rise buildings or arch dams, require deformation monitoring. Here is where area-wise deformation analysis covers the gap by implying measurement methods capable of remotely measuring a large area of the observed object [4]. To gain an impression of recent applications, the reader is referred to [5–7]. One recent method is Terrestrial Laser Scanning (TLS). Terrestrial Laser Scanners (TLSs) are active multi-sensor systems used to measure the three-dimensional geometry of a given surrounding within a certain range (cf. [8,9]). Laser scanners got more precise, compact and affordable in the past 20 years [10], but neither instrument manufacturers nor the scientific communities have reached common ground in what concerns all TLS influencing error sources. This is commonly known as the TLS error budget or TLS stochastic model; which is currently still unsatisfactory [11]. Neuner et al. [12] give an overview of the available point cloud modeling methods used in engineering geodesy together with their stochastic models and state that none of them is established. Generally, a stochastic model is a mathematical model that describes real-life phenomena that are characterized by the presence of uncertainty [13]. In any case of direct and indirect measurements [14], the stochastic model can be expressed by a variance-covariance matrix (VCM) [15]. If knowledge about the existing correlations between all observations is missing, the VCM is reduced to a diagonal matrix poorly resembling the complex nature of all the error sources (cf. [11]). This consequently leads to possibly wrong decisions in the TLS deformation analysis [16] or inappropriate estimations of a specific surface. (cf. [7]).

To overcome this issue, the Elementary Error Model (EEM) can be used to define the stochastic model of TLS observations in form of a VCM that considers correlations. Previous work of Kauker and Schwieger [17] sets the foundation of applying the EEM on TLS measurements. To that point, the EEM model was applied on a TLS of panoramic type [9] and the atmospheric elementary errors were considered functional correlating. Continuing this line of work, the current contribution introduces a model for long-range hybrid type TLSs and classifies the atmospheric elementary errors as stochastic correlating for the first time. The latter is possible due to derived correlations between atmospheric parameters in the research area. Results are shown on airside point clouds of the Kops arch dam in Vorarlberg, Austria.

In the second section of this paper, the EEM theory is reviewed for comprehension. Section 3 describes the application of EEM on a Riegl VZ-2000 hybrid TLS (RIEGL Laser Measurement Systems GmbH, Horn, Austria) together with meteorological elementary errors and their influences on the distance measurements and vertical angles. The study case and outcomes are presented in Sections 4 and 5 concludes this contribution.

2. Elementary Error Model Theory

2.1. General Remarks about Stochastic Models

The purpose of a stochastic model is to describe the statistical properties of variables [18]. There are many possibilities of describing the propagation of uncertainty of these variables. Out of these, some are most commonly used in measurements; these are: Guide to the Expression of Uncertainty in Measurement (GUM) [19], Monte Carlo Method (MCM) [20] and the variance covariance propagation law (cf. [21]). Only the last two will be briefly discussed with regard to the assumed models. On one side, in the MCM n random variables are numerically processed without having any knowledge about neither the linear/non-linear nature of the random variables nor their statistical distribution. Based on the outcomes, the statistical distribution is derived with corresponding parameters such as expected value, standard deviation, skewness and kurtosis. One disadvantage is that the model is computed n times, which increases computation time drastically. For more details the reader is referred to [20,22]. On the other side, variance covariance propagation law assumes normal distributed random values and linear or linearized models. Outcomes are likewise normally distributed and the statistical parameters are completely described by the expected value and standard deviation. This is an advantageous method, since the linear or linearized functional model is only computed once [18], therefore reducing computation time. It is also the main reason of adopting it for the EEM of TLS measurements, where the observations number easily reaches a few hundred thousand or a few million. To support this hypothesis, Aichinger and Schwieger [23] proved after using MCM for TLS observations for different scanning configurations that in 90% of the cases, linear models can be assumed with a significance level of $\alpha = 0.003$. Therefore, assuming a linear model for TLS observations is acceptable for most cases, even if observations have a non-linear nature. Regarding the numerical estimates introduced later, it is mentioned that no method of estimating the outcome's precision is currently used. This may be achieved in the future with the help of Variance Component Estimation

(VCE) based on [24,25], or a review of [26]. Our intention is to use sensitivity analysis (cf. [27,28]) and inspect how the input estimates influence the outcomes. All of these aspects will be prospectively presented in a different publication.

2.2. Elementary Error Theory

The general theory of the elementary error model was simultaneously defined by Hagen [29] and Bessel [30]. Later on, the model was elegantly presented by Pelzer [21] and extended by Schwieger [31]. Some of its applications can be found in exemplifying the error impact on several geodetic measurement methods like electronic distance measurement (EDM) instruments [32], GNSS observations [33] or recently TLS measurements [17].

According to the EEM theory, each realization of a measured random quantity differs from its expected value by a random deviation ε [31]. It is assumed that ε is comprised by the sum of countless, small elementary errors. Their absolute value is supposed to be equal and the probability of a positive and negative sign is likewise presumably equal [29]. The presumption of standard normal distribution of these errors is supported by an infinite number of elementary errors with infinitely small absolute values. Their impact on the observations can be modeled by using error vectors and influencing matrices. These matrices resemble the effect on the covariance matrix of observations. Three types of impacts are considered: non-correlating error vectors δ_k, functional correlating error vector ξ and stochastic correlating error vectors γ_h [31]. For each error type, corresponding influencing matrices are defined as follows: p matrices D_k for non-correlating errors, one matrix F for functional correlating errors and q matrices G_h for stochastic correlating errors. Therefore, the random deviation vector ε results as a sum of all elementary errors accordingly:

$$\varepsilon = \sum_{k=1}^{p} D_k \cdot \delta_k + F \cdot \xi + \sum_{h=1}^{q} G_h \cdot \gamma_h. \tag{1}$$

These influencing matrices have different structures depending on the elementary errors effects on the observations. Hereby, matrices D_k and G_h are symmetrical diagonal matrices because each elementary error of the non-correlating and stochastic correlating group influences exactly one measurement quantity functionally. The matrix F is fully populated because one functional correlating error may impact several measurement quantities [31]. Defining the functional relationships between observations $l_1 \ldots l_n$ and the elementary errors δ, ξ and γ allows the calculation of the partial derivatives that populate the influencing matrixes as follows:

$$D_k = \begin{bmatrix} \frac{\partial l_1}{\partial \delta_{1k}} & 0 & \cdots & 0 \\ 0 & \frac{\partial l_2}{\partial \delta_{2k}} & 0 & \vdots \\ \vdots & 0 & \ddots & \vdots \\ 0 & \cdots & \cdots & \frac{\partial l_n}{\partial \delta_{nk}} \end{bmatrix}, \quad F = \begin{bmatrix} \frac{\partial l_1}{\partial \xi_1} & \frac{\partial l_1}{\partial \xi_2} & \cdots & \frac{\partial l_1}{\partial \xi_m} \\ \frac{\partial l_2}{\partial \xi_1} & \frac{\partial l_2}{\partial \xi_2} & \cdots & \frac{\partial l_2}{\partial \xi_m} \\ \vdots & \vdots & \ddots & \vdots \\ \frac{\partial l_n}{\partial \xi_1} & \frac{\partial l_n}{\partial \xi_2} & \cdots & \frac{\partial l_n}{\partial \xi_m} \end{bmatrix}, \quad G_h = \begin{bmatrix} \frac{\partial l_1}{\partial \gamma_{1h}} & 0 & \cdots & 0 \\ 0 & \frac{\partial l_2}{\partial \gamma_{2h}} & 0 & \vdots \\ \vdots & 0 & \ddots & \vdots \\ 0 & \cdots & \cdots & \frac{\partial l_n}{\partial \gamma_{nh}} \end{bmatrix}. \tag{2}$$

Applying the law of propagation of variance on Equation (1) yields the so called "synthetic covariance matrix" Σ_{ll} which by definition has the following form [33]:

$$\Sigma_{ll} = \sum_{k=1}^{p} D_k \cdot \Sigma_{\delta\delta,k} \cdot D_k^T + F \cdot \Sigma_{\xi\xi} \cdot F^T + \sum_{h=1}^{q} G_h \cdot \Sigma_{\gamma\gamma,h} \cdot G_h^T, \tag{3}$$

where each covariance matrix of the elementary errors is defined and structured as shown below. The covariance matrices for non-correlating errors $\Sigma_{\delta\delta,k}$ and functional correlating errors $\Sigma_{\xi\xi}$ are diagonal matrices having the elementary error's variances on the main diagonal. As a result of the possible covariances of the stochastic correlating errors, the corresponding matrix $\Sigma_{\gamma\gamma,h}$ may be fully populated.

$$\Sigma_{\delta\delta,k} = \begin{bmatrix} \sigma_{1k}^2 & 0 & \cdots & 0 \\ 0 & \sigma_{2k}^2 & 0 & \vdots \\ \vdots & 0 & \ddots & \vdots \\ 0 & \cdots & \cdots & \sigma_{nk}^2 \end{bmatrix}, \Sigma_{\xi\xi} = \begin{bmatrix} \sigma_1^2 & 0 & \cdots & 0 \\ 0 & \sigma_2^2 & 0 & \vdots \\ \vdots & 0 & \ddots & \vdots \\ 0 & \cdots & \cdots & \sigma_m^2 \end{bmatrix}, \Sigma_{\gamma\gamma,h} = \begin{bmatrix} \sigma_{1h}^2 & \sigma_{12h} & \cdots & \sigma_{1nh} \\ \sigma_{12h} & \sigma_{2h}^2 & \cdots & \sigma_{2nh} \\ \vdots & \vdots & \ddots & \vdots \\ \sigma_{1nh} & \cdots & \cdots & \sigma_{nh}^2 \end{bmatrix}. \quad (4)$$

The challenging part is finding variances for all groups of errors and covariances for stochastic correlating errors. Correlations between the elementary errors are assumed to be zero. The variances may however be extracted from instrument manufacturers reports (cf. Section 3.2), empirical values (cf. Section 3.3) or an estimation based on the maximum error impact. In the last case, Pelzer [21] states that if the probability distribution is known, the standard deviation of an elementary error can be estimated with regard to its maximum error. Therefore, if a variable follows a rectangular distribution, the standard deviation is retrieved by multiplying the maximum error with 0.6. In case of a triangular distribution, multiplication is done by 0.4 and for normal distributions by 0.3. In what concerns the stochastic correlating group, values for the correlations must be supported by empirical values or literature. They represent stochastic relations for multi-dimensional normal distributed observations [17]. If the terms of Equation (3) are to be summed up, it can be seen that according to the matrices structures (see Equations (2) and (4)) the individual results are as follows: for non-correlating errors—a diagonal matrix, for functional and stochastic correlating errors—fully populated matrices. Thus, the synthetic variance-covariance matrix Σ_{ll} is also fully populated and illustrates the existing observation variances and covariances and indirectly their correlations.

3. Elementary Error Theory for Terrestrial Laser Scanners

3.1. Error Soruces in Terrestrial Laser Scanning

In order to apply the EEM on laser scanners, the error sources need to be identified and classified. As any measurement instrument, TLSs are realizations of an idealistic measurement system, therefore affected by physical manufacturing limitations. Even if the instrument itself would be hypothetically flawless, all measurements would be affected by the environment through which the electromagnetic waves are traveling (cf. [34]). Other error sources are related to the measured object properties such as surface material, roughness and color. These play an important role on the distance measurements and strongly depend on the used wavelength [35]. According to other authors (cf. [36]), scanning geometry is also considered an error source. Only instrumental and environmental error sources are treated in this contribution.

For a better understanding of how the TLS observations affect the coordinates (Figure 1), the mathematical relations between range (R), horizontal angle (λ), vertical angle (θ) and Cartesian coordinates (X, Y, Z) are described generically as follows:

$$\begin{aligned} X &= R \cdot \sin(\lambda) \cdot \cos(\theta), & R &= \sqrt{X^2 + Y^2 + Z^2}, \\ Y &= R \cdot \sin(\lambda) \cdot \sin(\theta), & \lambda &= \operatorname{atan}\left(\tfrac{X}{Y}\right), \\ Z &= R \cdot \cos(\theta), & \theta &= \operatorname{acos}\left(\tfrac{Z}{R}\right). \end{aligned} \quad (5)$$

3.2. Instrumental Elementary Errors

In comparison to the panorama TLSs architecture, the hybrid scanner architecture is less present in commercially available TLSs. This may be a reason for the reduced amount of scientific publications on calibration models for hybrid scanners. Even though it measures basically the same type of polar coordinates, calibration parameters (CPs) are of a more complex nature (cf. [37]). The most common example is the rotating polygon mirror used for deflecting the laser beam. On one side, the distance varies at each mirror position and on the other side, the EDM source is usually mounted with an offset from the rotation axis, not to mention that it may be intentionally tilted. Further on, a classification of the instrumental errors is necessary. Firstly, an explanation is given about how the errors are considered and afterwards numerical values are given.

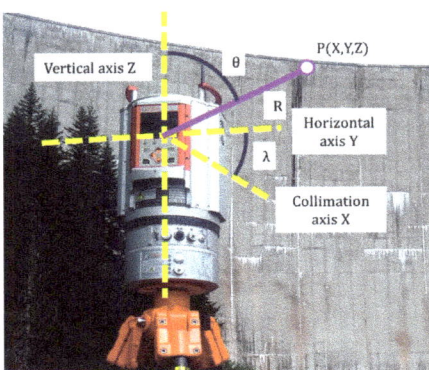

Figure 1. Example of Riegl Terrestrial Laser Scanner (TLS) and main axes.

To begin with, the non-correlating elementary errors are considered. These are measurement noise for angle and range measurements, which are not directly specified by the manufacturer. For range measurement, there is an entry for accuracy and one for precision. As defined by Riegl Laser Measurement Systems GmbH (Horn, Austria) [38], precision is the degree to which further measurements show the same result. If the definition of standard deviation is considered, it expresses how widely the random variable is spread out relative to the mean value of the sample [39]. Therefore, the given value for precision will be used as an indicator for instrument internal range noise at all measured ranges (see Table 1). For angle measurements, the data sheet of the instrument offers only "angle resolution" without further details. According to Wunderlich et al. [40], angular resolution can be interpreted as measurement precision (one sigma), therefore the same convention is used (see Table 1). The terms are generally presented in Equation (6) and their values are found in Table 1. Having this, the first term of Equation (2) and the first term of Equation (4) are now defined as follows:

$$\Sigma_{\delta\delta,k} = \begin{bmatrix} \sigma_\lambda^2 & 0 & 0 \\ 0 & \sigma_\theta^2 & 0 \\ 0 & 0 & \sigma_R^2 \end{bmatrix}, \; D = I. \quad (6)$$

Table 1. Classification of instrumental errors and their dimensions.

Type of Error	Parameter	Standard Deviation
Non-correlating errors	Range noise	$\sigma_R = 5$ mm
	Angle noise (λ, θ)	$\sigma_\lambda = 0.55$ mgon
		$\sigma_\theta = 1.66$ mgon
Functional correlating errors	a_0	$\sigma = 0.34$ mm
	a_1	$\sigma = 40$ ppm
	b_4	$\sigma = 3.18$ mgon
	b_6	$\sigma = 1.91$ mgon
	c_0	$\sigma = 1.08$ mgon
	c_1	$\sigma = 1.85$ mgon
	c_4	$\sigma = 0.64$ mgon

The influencing matrix D is the identity matrix, because no transformation from the coordinate space into observation space is needed at this moment. Only after the complete synthetic VCM is computed, a transformation based on Equation (5) is made from observation space to coordinate space.

Regarding the functional model of observations, a model defined by Lichti [41] and Lichti [42], later simplified by Schneider [43] is adopted. The latter applied it on a Riegl LMS-420i and could successfully improve the results after calibration. The simplification is mostly justified by the fact that not all of the CPs can be classified as significant after a calibration. Furthermore, if they are highly correlated they only reduce the validity of the model. Some of them are negligible, some are not determinable or separable, and therefore the used model is restricted to the minimum number of CPs identified as significant [37]. For more details about these parameters, the reader is advised to consult [37,43]. Following [43], the CPs for each observation can be defined as stated:

$$\Delta R = a_0 + a_1 R + a_2 R^2,$$
$$\Delta \lambda = b_1 \sec(\theta) + b_2 \tan(\theta) + b_3 \sin(\lambda) + b_4 \cos(\lambda) + \arcsin\left(\frac{b_5}{R}\right) + b_6 \sin(2\lambda) + b_7 \cos(2\lambda) + b_8 \cos(3\lambda), \quad (7)$$
$$\Delta \theta = c_0 + c_1 \sin(\theta) + c_2 \cos(\theta) + \arcsin(c_3/R) + c_4 \cos(3\lambda),$$

where a_0 is the zero point error, a_1 scale error, a_2 quadric scale error, b_1 collimation axis error, b_2 horizontal axis error, b_3 and b_4 first and second horizontal circle eccentricity, b_5 eccentricity of the collimation axis with respect to the vertical axis, b_6 and b_7 non-orthogonality of the plane containing the horizontal angle encoder and the vertical axis, b_8 empirical parameter for compensation of remaining systematic effects, c_0 vertical circle index error, c_1 and c_2 first and second vertical circle eccentricity, c_3 eccentricity of the collimation axis with respect to the trunnion axis, c_4 empirical parameter to model a sinusoidal errors function of the horizontal direction with period of 120° (cosine term).

Out of all CPs, only some of them have numerical values and were determined as significant after calibration by Schneider [43]. For the EEM, variances of the CPs are introduced in the middle term of Equation (4) and the F matrix contains the partial derivatives of Equation (7). The values for the variances are presented in Table 1 with adopted dimensions for the Riegl VZ-2000 scanner. Further investigations on the hybrid scanner architecture are in progress based on the foundations set in [44].

3.3. Meteorological Elementary Errors

3.3.1. Influences on the Distance Measurement

Similar to EDM of total stations, distance measurements in TLS, are influenced by air temperature and air pressure; in any case for long distances. Partial water vapor pressure is intentionally neglected due to its small influence. Most TLSs use near-infrared light for measuring distances. As known, the speed of light traveling through the atmosphere's different layers is diminished in comparison to the speed of light in vacuum. The atmospheric correction increases proportionally with the measured distance [34]. In case of ranges up to 200 m, these corrections may be neglected, but it cannot be neglected for long range scanners (e.g., Riegl VZ-2000) that measures up to 2050 m. According to manufacturer's specifications, the Riegl scanner has an atmospheric correction model implemented in the instrument, meaning that distances are corrected based on the introduced parameters for temperature, pressure and relative humidity. Information of how this happens can be taken from the RiSCAN Pro software documentation [45] and further inspected in the IAG 1999 resolutions [46]. The authors retain from explaining the whole process of retrieving the influencing coefficients for distance measurement and directly give the formula implemented in the EEM:

$$\Delta n \cdot 10^{-6} = -0.93 \cdot \Delta t + 0.27 \cdot \Delta p, \Delta R = -R \cdot \Delta n, \quad (8)$$

where Δn is the the change of the group refractive index of light, Δt is change in temperature (°C) and Δp is change in pressure (hPa). Finally, the change in range ΔR is given. Note that these parameters are calculated for a mean atmosphere of 17 °C, 1000 hPa pressure and a wavelength of λ = 1550 nm.

Interpreting this in terms of parts per million (ppm) depending on the two atmospheric parameters in standard conditions, a change in t of 1 °C affects the distance and refractive index by 0.93 ppm, a change in air pressure of 10 hPa yields a −2.7 ppm correction on the distance. For more details about this topic, the reader can consult [47] or [34].

3.3.2. Influences on the Vertical Angle Measurement

In addition to the effects on distance measurements of any electro-optical measurement, atmospheric refraction also influences the vertical angle measurements. This effect causes image scintillation, often obvious in its extreme case when temperature gradients near the ground are high (e.g., in the desert or on a highway in hot summer days). This mostly affects angle measurements and is likewise important in geodesy, receiving much attention in transferring heights by trigonometric leveling. Nevertheless, this effect also occurs in TLS measurements and has been empirically studied by Friedli et al. [48]. The reader is advised to consult this work for understanding how refraction angles can be determined with the aid of reference values from total station measurements.

Figure 2 denotes the effects of atmospheric refraction out of which the refraction angle correction $\delta/2$ is of further interest. This angle is given between the expected wave path and apparent line of sight also called tangent to refracted wave path. For more details about how $\delta/2$ is deduced, refer to [49]. There are different ways of expressing the refraction angle correction, but only one has been chosen based on its simplicity and implemented in the EEM. The choice is not relevant in case of stochastic modeling. The corrected vertical angle can be therefore computed [49]:

$$\theta = \theta' + \frac{\delta}{2}, \frac{\delta}{2} = \frac{R}{2 \cdot E_R} \cdot k \cdot \rho, \qquad (9)$$

where θ is the corrected vertical angle, θ' the measured vertical angle, R measured range, E_R Earth's middle radius (6381 km), k refraction coefficient and ρ conversion constant between angle measurement units (degrees or grads) and radians. The coefficient of refraction k is usually needed to account for the curved light path from one point to another. It is defined as a ratio between the Earth radius and the radius of the line of sight which is mostly convex [50]. Very often, the Gaussian value of $k = +0.13$ is used by default as a setting for total station measurements, hoping that it holds true for most applications [51]. Nevertheless, k strongly varies throughout the day and is directly dependent on the temperature gradient $\partial T/\partial Z$ (K/m) (cf. [52]). If the refraction coefficient of a particular point is of interest, the local refraction coefficient k_{loc} is given as a function dependent on temperature, pressure and the local temperature gradient (cf. [49,52]):

$$k_{loc} = 503 \cdot \frac{p}{T^2} \cdot \left(0.0343 + \frac{\partial T}{\partial Z} \right), \qquad (10)$$

where p is pressure (hPa), T is temperature in (K) and $\partial T/\partial Z$ (K/m) is the temperature gradient at a certain point. The term k_{loc} is used instead of an average k in Equation (9) for further purposes. As noticed in Equation (10), temperature gradient strongly determines the size of the local refraction coefficient; hence, its variation from ground level up to 100 m above the ground, as relevant for the later given examples, will be discussed. This is defined in meteorology or climate research under the name of micro- and local climate [53]. Hirt et al. [54] use the terms higher, intermediate and lower atmosphere to define the variation of the vertical temperature gradient (VTG) within a given range. By higher atmosphere, the layers from 100 m and above the ground surface are addressed. The VTG in this part of the troposphere has values around −0.006 K/m and is fairly independent of the Earth's surface temperature [54]. The next layer, the intermediate atmosphere between 20–30 m and 100 m is weakly influenced by the ground temperature and has an average value for the VTG of −0.01 K/m. This is where the refraction coefficient has an average value of +0.15 and it is also the layer to which the Gaussian value is most appropriate. Going a level lower, the first layer, considered the lower atmosphere is where ground temperature reaches its maximum influence on the VTG. Several studies, summarized in [54], showed variations of the refraction coefficient between −3.5 and 3.5. Noteworthy are the empirical findings of Hennes [55] in which the local refraction coefficient reaches values of −2.9 (from a VTG of −0.5 K/m) leading to a concave curvature of the light path, contrasting the common belief about the chord being convex in almost all cases. Nevertheless, a less drastic value of −0.2 K/m is used in the current study, resembling an average value for this layer.

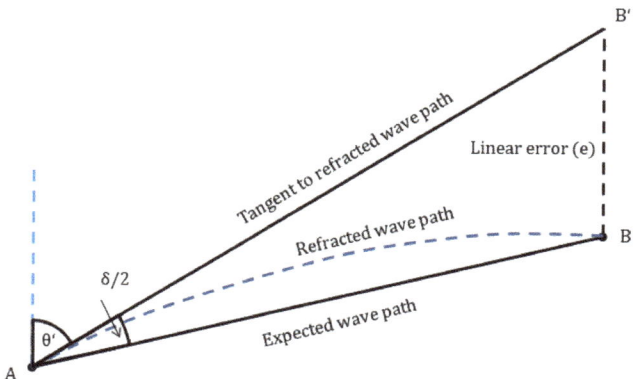

Figure 2. The effects of atmospheric refraction (after [3]).

Similar as in Section 3.3.1, the influencing coefficients are determined after computing the partial derivatives of Equations (9) and (10). Therefore, numeric values have been exemplary computed with the same conditions as stated before ($t = 17\,°C$, $p = 1000$ hPa, VTG = -0.01 K/m) at a distance of 1000 m. The change in the measured vertical angle (in radians) is given by:

$$\Delta\theta \cdot 10^{-6} = -0.08 \cdot \Delta t + 0.01 \cdot \Delta p + 468.17 \cdot \Delta VGT, \qquad (11)$$

where $\Delta\theta$ is the the change in the measured vertical angle, Δt is change in temperature (°C), Δp is change in pressure (hPa) and ΔVGT is the change in vertical temperature gradient. In other words, a change in temperature of 10 °C affects the vertical angle by −0.8 μrad (−0.05 mgon), a change in air pressure of 10 hPa affects the vertical angle by 0.1 μrad (+0.006 mgon) and the most significant factor, a change in the VGT of 1 K/m results in a change of the angle with 468.17 μrad (29.8 mgon). This is not to be confused with the systematic effect of the refraction angle correction $\delta/2$. For comprehension, for the above stated conditions and at 1000 m, $\delta/2$ has a value of 0.7 mgon, which leads to a value of the linear error e = 11.4 mm. The intention is not to correct these systematic effects, but to show how varying temperature and pressure influence the error of position.

Although often not considered, air pressure also follows a gradient. This is less variable than the VGT and according to the Deutscher Wetterdienst Lexikon [56], the pressure gradient throughout the mentioned atmospheric layers is $\delta p/\delta z = -0.125$ hPa. This information will also be further used.

All these layer definitions and given values are adopted further in Section 3.4 to derive the necessary variances and covariances needed in the EEM.

3.4. Atmospheric Errors as Stochastic Correlating Errors

In most terrestrial precision measurements, if the atmospheric parameters temperature, pressure and relative humidity are needed, they are measured at the station point and in some cases near the observed object or second station point. For corrections, an averaged value of these parameters is used in most cases. This may hold true for airborne laser scanning, where the average between aircraft and ground temperature is sufficient [57], but in TLS, the situation changes. In addition to this, it was shown (cf. [48,54]) that even within a short time span, these may present strong variations and there is no straightforward method of correcting the measurement values dependent on these temporal variations. For this reason, modeling the impacts on the observations needs to be done stochastically. To do so, a VCM of the varying terms air temperature (t), air pressure (p) and VGT (g) is needed. The challenge is to fully populate the VCM $\Sigma_{\gamma\gamma,h}$ from Equation (4) so that existing correlations between all elementary errors are known. This will have the general form in case of t, p and g (only upper diagonal presented) as follows:

$$\Sigma_{\gamma\gamma,atm} = \begin{bmatrix} s_{t1}^2 & s_{t1p1} & s_{t1g1} & s_{t1t2} & s_{t1p2} & s_{t1g2} & \cdots & s_{t1tm} & s_{t1pm} & s_{t1gm} \\ & s_{p1}^2 & s_{p1g1} & s_{p1t2} & s_{p1p2} & s_{p1g2} & \cdots & s_{p1tm} & s_{p1pm} & s_{p1gm} \\ & & s_{g1}^2 & s_{g1t2} & s_{g1p2} & s_{g1g2} & \cdots & s_{g1tm} & s_{g1pm} & s_{g1gm} \\ & & & s_{t2}^2 & s_{t2p2} & s_{t2p2} & \cdots & s_{t2tm} & s_{t2pm} & s_{t2gm} \\ & & & & s_{p2}^2 & s_{p2g2} & \cdots & \vdots & \vdots & \vdots \\ & & & & & s_{g2}^2 & \cdots & & & \\ & & & & & & \ddots & \vdots & \vdots & \vdots \\ & & & & & & & s_{tm}^2 & s_{tmpm} & s_{tmgm} \\ & & & & & & & & s_{tm}^2 & s_{pmgm} \\ & & & & & & & & & s_{gm}^2 \end{bmatrix} \quad (12)$$

The main diagonal is not difficult to fill in, according to what will be explained next, but the rest of the elements on all the upper and implicitly lower part of the same matrix are actually the challenge. To overcome this, correlation coefficients between t, p and g are computed for the given spatial distribution of all observations. This will be further on explained.

As in any terrestrial electro-optical measurement, in TLS observations light travels from the instrument to the measured object and back. Due to varying atmospheric conditions, it will be perturbed throughout the whole path and in order to evaluate how air temperature, air pressure and VTG vary along the path, pre-knowledge about these parameters (cf. Section 3.3.2) is used in a combined way with spatial information. Suppose a laser scanner is stationed at a certain distance near a tall object and observations are possible from the base to the tip of that object. If a rough digital terrain model (DTM) of the area is available, then the local topography is known, which further allows a classification of the VGT depending on how the topography varies. Simply explained, the limits of the gradient layers can be defined as surfaces with an offset from ground level according to how meteorologists have defined these limits (Figure 3 left). The yellow surface defines the separating layer at about 25 m between the lower and intermediate atmosphere; the red layer is the separation between intermediate and higher atmosphere at about 100 m above the ground. In order to have a better overview of the further steps, a vertical section is selected and exemplified in Figure 3 right.

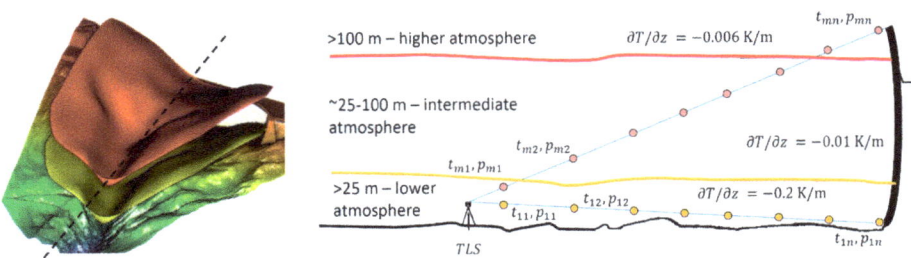

Figure 3. (**Left**) Spatial separation of vertical temperature gradient (VGT) layers; (**right**) one section of the spatial VGT model.

It is necessary to roughly know the position of the scanner on the DTM. This is often referred to as georeferencing, but in this case the accuracy of the scanner's position is not of high importance, therefore an approximation suffices. In most cases, the air temperature and air pressure is measured near the laser scanner, usually at the instrument height. According to the situation depicted above, this is true for temperature and pressure only near the laser scanner, but the interest is in gaining information about the atmospheric parameters along the whole measurement path. Therefore, in the next step the observation lines are reconstructed in relation to the scanner position on the DTM. This directly shows which observation line passes through which atmospheric layer. Only two of them are depicted by the blue lines in Figure 3 right, but the same principle applies for all the rest. Further on, a series of points along these lines are selected; denoted by the yellow and pink circles. Point spacing along the observation line may be done subjectively; but a uniform distribution between scanner and object is suggested for representative results. Each of these points receive values for t, p and g, determined according to their position in space and in relation to the measured atmospheric parameters at the station point denoted by "TLS" in Figure 3 right. For example, along the lowest observation line, each yellow point will receive a value starting from t_{11}, p_{11}, g_{11} to t_{1n}, p_{1n}, g_{1n}. The individual values are extrapolated according to the individual position. If more measurements of the atmospheric parameters at other positions within the DTM are available, they can be considered within the extrapolation processes. This applies for all the other observation lines, until the highest one is reached. In this case, the pink points (Figure 3 right) receive t_{m1}, p_{m1}, g_{m1} to t_{mn}, p_{mn}, g_{mn}. Notice that the points are chosen on the same vertical line, fact that will be explained later. Having series of values for all three parameters along all observation lines allows the computation of variances and covariances along and between each observation line according to:

$$s_a^2 = \frac{1}{n-1} \cdot \sum_{i=1}^{n} (a_i - \overline{a_i})^2,$$
$$s_{a_i a_j} = \frac{1}{n-1} \cdot \sum_{i,j=1}^{n} (a_i - \overline{a_i}) \cdot (a_j - \overline{a_j}), \tag{13}$$

where s_a^2 is the empirical standard deviation computed for each of the three atmospheric parameters. The value a_i and a_j are replaced consecutively by t_i, p_i, g_i and n is the number of points along each observation line, $s_{a_i a_j}$ is the empirical covariance between pairs of the three parameters. To exemplify this, the general VCM $\Sigma_{\gamma\gamma,atm}$ for one observation line has the following form:

$$\Sigma_{\gamma\gamma,atm} = \begin{bmatrix} s_t^2 & s_{tp} & s_{tg} \\ s_{tp} & s_p^2 & s_{pg} \\ s_{tg} & s_{pg} & s_g^2 \end{bmatrix}. \tag{14}$$

Out of this VCM, a correlation matrix is computed, out of which correlation coefficients are extracted. For example, along each line, the correlation matrix R_{atm} is obtained and contains the following correlation coefficients r_{tp}, r_{tg}, r_{pg}. This is valid for all lines up to the n-th observation.

$$R_{atm} = \frac{1}{\sqrt{diag(\Sigma_{\gamma\gamma,atm})}} \cdot \Sigma_{\gamma\gamma,atm} \cdot \frac{1}{\sqrt{diag(\Sigma_{\gamma\gamma,atm})}} = \begin{bmatrix} 1 & r_{tp} & r_{tg} \\ r_{tp} & 1 & r_{pg} \\ r_{tg} & r_{pg} & 1 \end{bmatrix}. \tag{15}$$

This is valid along observation lines and helps at filling in the block matrices on the main diagonal of VCM from Equation (12) with submatrices like in Equation (14). For all other elements of $\Sigma_{\gamma\gamma,atm}$ the covariances are computed with the help of the correlation coefficients according to:

$$s_{ij} = r_{ij} \cdot s_i \cdot s_j. \tag{16}$$

The values for s_i and s_j are computed along each observation line with the help of Equation (13). Following Equation (15), a set of values for r_{tp}, r_{tg}, r_{pg} is obtained. In addition to these, the correlation coefficients of the same parameters (t–t, p–p and g–g) between the observation lines have to be determined r_{tt}, r_{pp}, r_{gg} and to accomplish this, VCMs and correlation matrices are computed between each parameter of the observation lines (e.g., t_{11} and t_{m1}). This is also the reason why it is relevant to have the yellow and pink points on the same vertical line. In this way values for each of the atmospheric parameters are treated as series of values for the same dimension, in this case vertical direction. A drawback of this proposal is that a number of n TLS observations leads to a number of $n!/(n-3)!$ permutations of correlation coefficients. This means that for e.g., 200 values taken three times (t, p, g) one would obtain a number of 7,880,400 correlation coefficients that need to be properly arranged in $\Sigma_{\gamma\gamma,atm}$. This is currently not achievable due to technical reasons for TLS observations where the number of observations easily reaches a few million. Therefore, one generic value is taken for each of the coefficients r_{tp}, r_{tg}, r_{pg} and r_{tt}, r_{pp}, r_{gg}. Numeric values are computed between the lowest and highest observation line taken from the vertical section as denoted in Figure 3 right. Finally, the individual values for the covariances are computed based on Equation (16) and then introduced in $\Sigma_{\gamma\gamma,atm}$ like in Equation (12). Returning to the EEM, now that the matrix $\Sigma_{\gamma\gamma,atm}$ is available, the influencing matrix G_h from Equation (2) must be properly filled. The complete matrix is a block matrix that has the partial derivatives of the observations with respect to t, p and g, as presented in Equation (17).

$$G_{atm} = \begin{bmatrix} G_1 & 0 & \cdots & 0 \\ 0 & G_2 & 0 & \vdots \\ \vdots & 0 & \ddots & \vdots \\ 0 & \cdots & \cdots & G_n \end{bmatrix}, \quad G_i = \begin{bmatrix} \frac{\partial \lambda}{\partial t} & \frac{\partial \lambda}{\partial p} & \frac{\partial \lambda}{\partial g} \\ \frac{\partial \theta}{\partial t} & \frac{\partial \theta}{\partial p} & \frac{\partial \theta}{\partial g} \\ \frac{\partial R}{\partial t} & \frac{\partial R}{\partial p} & \frac{\partial R}{\partial g} \end{bmatrix}, \tag{17}$$

where $i = 1 \ldots n$ and G_1 to G_n are block matrices that have the partial derivatives as shown above. Effects on the horizontal angles have not been discussed and are not considered in this model. Therefore, the first line of each block g will be filled with 0. The second line includes the coefficients presented in Section 3.3.2 in Equation (11) and this is the only line that has influencing values for all variations in the fully populated VCM (Equation (12)); the last line of g has the influencing values presented in Section 3.3.1, Equation (8) with the last element 0. The numerical values must be computed with regard to given atmospheric conditions and at a given range for each situation. Finally, the last term of the synthetic VCM can be computed and therefore the influences of instrumental and atmospheric parameters can be combined. In the upcoming section, the whole methodology presented until now will be applied for TLS point clouds of an arch dam.

4. Study Case: Arch Dam Kops

The Kops water dam is a storage concrete dam built between 1962 and 1969 in Vorarlberg, Austria. It is considered a hybrid dam made out of a gravity dam and an arch dam with artificial counterfort or abutment. It retains a volume of almost 43 million m³ of water, thus creating the 1 km² "Kopssee" lake [58]. Only measurements of the downstream (airside) arch dam are considered. For this reason, its dimensions are mentioned to give a general impression. The crown spans over 400 m, its height is 122 m from foundation to crest and has a crest width of 6 m. Between 29 July and 2 August 2019, a first measurement campaign of the Kops dam took place and part of the results from another type of laser scanner are presented by Kerekes and Schwieger [59]. Further on, the EEM is applied on point clouds acquired with the Riegl VZ-2000 from varying positions.

In order to apply the EEM for meteorological elementary errors as described in Section 3.4, a DTM for the area of interest was cordially made available by the "Landesamt für Vermessung und Geoinformation" Land Vorarlberg, Austria. TLS Point clouds were acquired from four different station points. Figure 4 shows the distribution of these on the DTM together with an example of a vertical section plane out of which temperature and pressure are extracted. Results will be presented for all four station points (S1–S4).

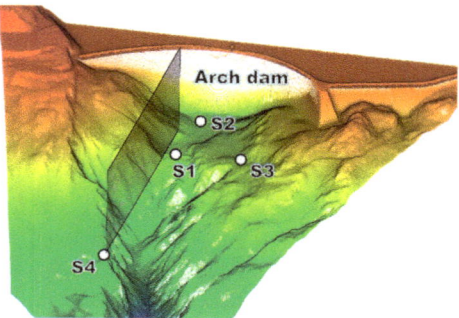

Figure 4. TLS station points and vertical section on digital terrain model (DTM) (Source for DTM: LiDAR Data obtained from Land Vorarlberg, Austria—data.vorarlberg.gv.at).

In order to have an overview about the varying scanning configurations and atmospheric conditions, Table 2 summarizes all the relevant parameters.

Table 2. Overview of scanning configurations and atmospheric conditions.

Station Point	Shortest Horizontal Distance from Scanner to Dam	Mean Distance to Dam	Weather Conditions
S1	180 [m]	194 [m]	$t = 16\,°C, p = 941$ mbar
S2	88 [m]	119 [m]	$t = 15\,°C, p = 942$ mbar
S3	196 [m]	203 [m]	$t = 23\,°C, p = 832$ mbar
S4	456 [m]	466 [m]	$t = 24\,°C, p = 840$ mbar

Considering the harsh local topography with steep slopes and vegetation, only the four station points depicted in Figure 4 were measurable within reasonable time, effort and coverages of the dam airside. With exception of S4, all other point clouds cover more than 80% of the airside surface. Weather recordings were made at each station point at the instrument height approx. 2 m above ground height with the Greisinger precision Thermo-, Barometer GTD1100 (Greisinger GmbH, Regenstauf, Germany). According to the technical specifications, air temperature is measured with an accuracy of +/−1% of the reading in the interval −10 °C to +50 °C and air pressure with +/−1.5 hPa in the interval of 750 hPa to 1100 hPa [60]. These accuracies are considered in the process of determining t and p. As regards the VGT, an empirical value for the uncertainty can be found in [55] for an Alpine region:

$$\sigma_{VGT} = \sqrt{\frac{1}{12} \cdot (VGT_{max} - VGT_{min})^2}, \qquad (18)$$

where a value for $\sigma_{VGT} = 0.25$ K/m in case of the Alpine region in the lower atmosphere is given and VGT_{max} and VGT_{min} are the upper and lower numerical recorded values for the VGT [55].

In case of the other two layers, no empirical values for variances have been found to the best of the author's knowledge. Due to this, values are obtained by multiplying the VGT value with 0.3 following the explanation in Section 2.1; these variations are likewise considered in determining the values for t.

For all station points the same methodology is applied, but the vertical section is only visualized in the case of S4 (Figure 5). The authors consider this case the most interesting since the distance to the dam is the longest and observations pass through different atmospheric layers more than once.

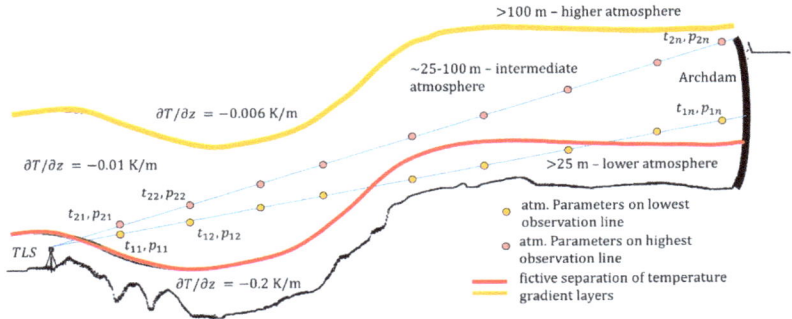

Figure 5. Vertical section through the terrain and temperature gradient fictive separation.

Almost half of the observation lines in this profile pass through the lower layer twice, meaning that variances of temperature and VGT affect the lowest points in the point cloud more than the ones obtained from observations that travel through a more stable atmospheric layer. This is confirmed further when analyzing the error of position in the point cloud. To have an overview of all variances and spatial correlation coefficients, Table 3 presents them for all station points.

Table 3. Overview of variances and spatial correlation coefficients for atmospheric elementary errors.

Station Point	Temperature Variance from Lowest to Highest Line ($\sigma_{t1}\ldots\sigma_{tn}$)	Temperature Variance from Lowest to Highest Line ($\sigma_{p1}\ldots\sigma_{pn}$)	Temperature Variance from Lowest to Highest Line ($\sigma_{g1}\ldots\sigma_{gn}$)	Correlation Coefficients					
				r_{tt}	r_{pp}	r_{gg}	r_{tp}	r_{tg}	r_{pg}
S1	0.28 … 2.36 °C	1.55 … 3.36 hPa	0.25 … 0.17 K/m	0.27	0.08	0.44	0.19	0.27	0.10
S2	0.17 … 1.13 °C	1.53 … 3.44 hPa	0.24 … 0.16 K/m	0.28	0.23	0.33	0.49	0.18	0.36
S3	1.94 … 1.78 °C	1.54 … 3.14 hPa	0.25 … 0.17 K/m	0.27	0.24	0.20	0.59	0.43	0.35
S4	1.88 … 1.61 °C	4.01 … 4.78 hPa	0.11 … 0.03 K/m	0.48	0.80	0.31	0.31	0.63	0.47

The last pairs of correlation coefficients r_{pt}, r_{gt}, r_{gp} are not in the table because they have the same value with r_{tp}, r_{tg}, r_{pg}. Note that these values resemble the spatial correlations only. The subject of temporal correlations will be addressed in a future publication after a second measurement epoch is available. The variances and correlation coefficients from Table 3 are used to finally create a VCM $\Sigma_{\gamma\gamma,atm}$ for the atmospheric elementary errors for each station point. In case of the instrumental errors, all the values are the same as stated in Table 1 since the same instrument was used in all station points.

The EEM is implemented in Matlab—MathWorks and currently limited to handling VCMs having sizes of up to 21,000 × 21,000 cells. More details about this can be found in [59]. Before applying the EEM on the point clouds, only points on the dam are selected and a subsampling is done. Consequently, the complete point cloud contains points on the dam airside with a distance of 1.5 m between them. Additional to this, vertical sections on the dam are analyzed since much attention was accorded to how atmospheric parameters vary along vertical profiles. The point spacing in the section is denser with an average distance of 15 cm between the points. This is done due to technical restriction mentioned above.

Coordinates (X, Y, Z) are considered instead of observations (R, λ, θ), because the VCM will be used for estimating the geometric primitives of a B-Spline surface in the future [59]. Therefore, the synthetic VCM is computed in observation space and then transformed with the help of equations 5 to coordinate space. Results are presented with regard to the error of position, spatial correlations along a vertical line chosen to be as long as possible along the scan (Figure 6) and contribution of all variances to the error budget in case of a single point.

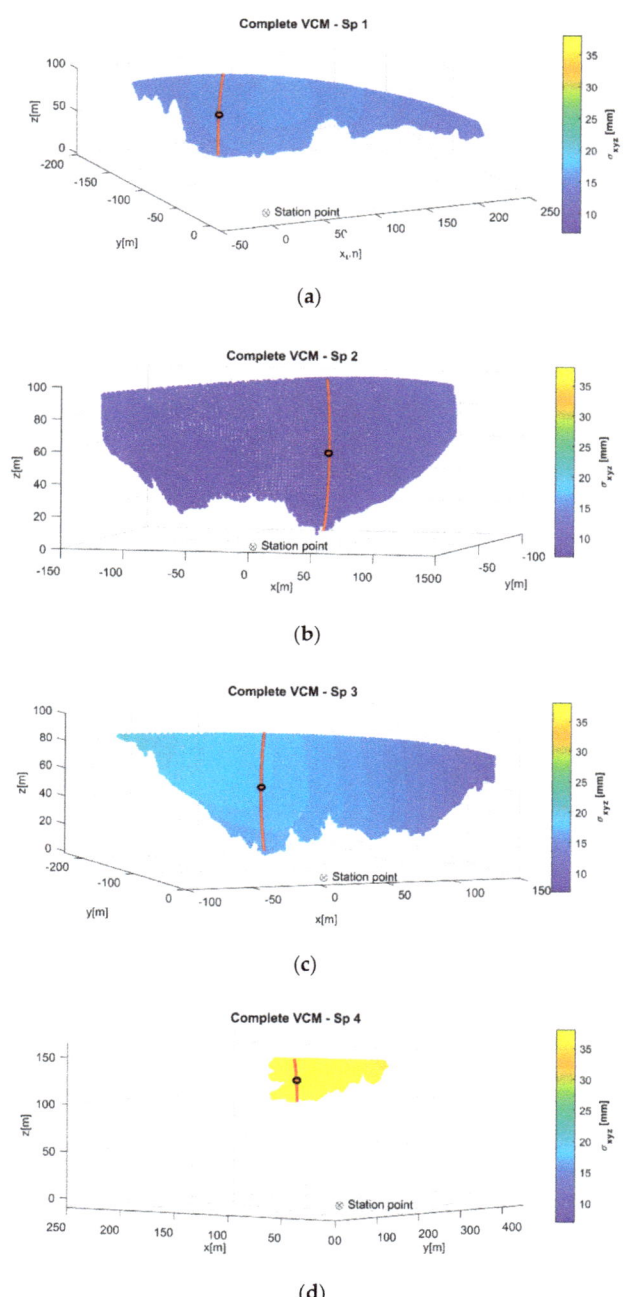

Figure 6. Example of error of position distribution on point clouds, selected vertical sections (red lines) and selected points (red circles). Scanning station points are (**a**) S1, (**b**) S2, (**c**) S3, (**d**) S4.

4.1. Error of Position

The main diagonal of the synthetic VCM in coordinate space contains the variances for each point in the point cloud. The error of position is computed according to [21]:

$$\sigma_{xyz\_i} = \sqrt{\sigma_{xi}^2 + \sigma_{yi}^2 + \sigma_{zi}^2}. \tag{19}$$

Points are visualized with respect to their position in space (local coordinate system) relative to the laser scanner position (0, 0, 0) and the dimension of the error of position is given on a color scale. Figure 6 exemplary shows the results obtained from all four station points, the selected vertical section and a point for which the percentage contribution of the variances is given later on.

Here the coverage of the air side can be seen for the first time. As expected from Figure 4, S1 to S3 cover most of the dam's airside, whilst S4 (Figure 6d) has the smallest coverage due to height difference and vegetation. Outcomes led to average errors of position, as follows: S1 – $\bar{\sigma}_{xyz}$ = 15.2 mm, S2 – $\bar{\sigma}_{xyz}$ = 9.6 mm, S3 – $\bar{\sigma}_{xyz}$ = 16.4 mm and S4 – $\bar{\sigma}_{xyz}$ = 36.5 mm. A common color scale was chosen to maintain comparability; this is why the reader is asked to consult the digital version of the paper. At a first glance, it can be seen in which way the errors of position are distance dependent, with the smallest values for S2 (Figure 6b) which is the nearest point and the biggest value for S4 (Figure 6d) at a distance of 466 m. Both S1 and S3 (Figure 6a,c) present similar results due to the similar measurement configuration. At a closer inspection of the scan from S4, it can be seen how the error of position decreases with height, reaching a minimum at the crest (dark yellow). The lowest part has the highest values for the error of position (bright yellow). This resembles the smaller variation of the VGT (see Table 3) for TLS observations that pass through more stable atmospheric layers.

4.2. Spatial Correlations Along Vertical Sections

Existing correlations can be analyzed after computing a correlation matrix based on the VCM (Equation (15)). Generally, high correlations are an indication for high variances. Previous publications of the EEM for TLS topic (cf. [17]) treated atmospheric elementary errors as functional correlating. In this contribution, atmospheric elementary errors were treated stochastic correlating (cf. Section 3.4) for the first time. This leads to different spatial correlations than have not been discussed before. For this reason, special attention is offered to the stochastic correlating errors and presented in parallel with the ones obtained from the complete VCM where instrumental errors also influence the results. For each station point, one vertical section has been selected and spatial correlations are presented between the lowest point of the section and all other. The analysis is made only for the height coordinates Z. Results do not change if observations would be analyzed instead. The reason for choosing only one vertical section is that emphasis is put on how the stochastic correlating errors influence the correlations and height error of position in comparison with the complete VCM of the same station point. These cases are shown in parallel. In the first case, the EEM considers instrumental and atmospheric errors (Figure 6 left side) and in the second only atmospheric errors are considered in the EEM (Figure 6 right side).

Analyzing all correlations when instrumental and atmospheric elementary errors are considered, values are in almost all cases higher than 0.5 (Figure 7a,c,d) and present a linear decrease with increasing height. The same effect is noticed with the standard deviations of the heights that remain at approximately the same level. The exception to this is S2 (Figure 7b) where the correlations decrease with height and standard deviation increases. This is the station point with the smallest distance to the dam; therefore, an explanation for this effect is analyzed in the next section. Correlations in case of the atmospheric errors all show a linear behavior (Figure 7e–h), but at different levels. As maybe presumed, the standard deviations are very small at these distances and given level of variation for the atmospheric parameters, but the most interesting finding is at S4 (Figure 7h) where the decrease in standard deviation of height is obvious, explaining the presumption made in Section 4.1 that the upper observations travel through more stable layers of the atmosphere and are less affected by variations.

This may lead to the thought that scans for e.g., objects over a valley are more reliable than those acquired from parallel to the ground level. This is partly true, since VGT may also be stable for a short period of time at ground level; therefore, this issue strongly depends on the topography and local conditions. If the conditions in Table 2 are reviewed, it can be seen that similar atmospheric conditions do not necessarily lead to a similar level of correlation, for example point clouds from S2 and S4 where acquired under differing conditions, but led to a similar level of correlation for the atmospheric elementary errors.

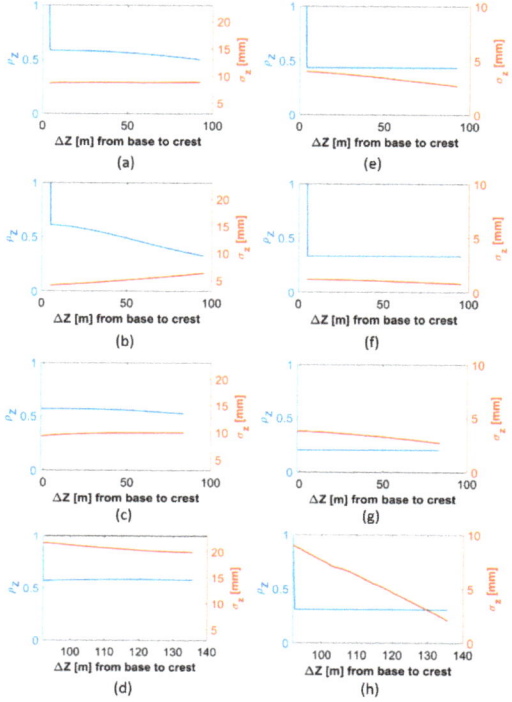

Figure 7. Left: spatial correlations, standard deviations (height)—complete synthetic VCM (**a**) S1, (**b**) S2, (**c**) S3, (**d**) S4; right: spatial correlations, standard deviations (height)—synthetic VCM with atmospheric elementary errors only (**e**) S1, (**f**) S2, (**g**) S3, (**h**) S4.

4.3. Contribution of the Elementary Errors Variances to the Overall Error Budget

The points depicted in Figure 6 (red circles) are chosen to analyze the contribution of variances to the whole error budget. Note that here, all dimensions (X, Y, Z) are considered and not only Z as in the section before. For the first three station points, they are approximately at the same level. In case of S4, this is not possible since that part of the dam cannot be scanned.

In all situations, instrumental errors make up the majority of the error budget (Figure 8). The parameters b_4—horizontal circle eccentricity, a_1—scale error and c_1—horizontal circle eccentricity comprise in all cases more than 50% of the error budget. It can also be seen that some instrumental errors are negligible (e.g., a_0) since they remain under the 1% quote. In the previous section, the standard deviations for Z are affected by instrumental errors of the vertical angle encoder (c_0, c_1, c_4) and vertical angle noise σ_θ. It is also seen how this phenomenon is almost independent of distance and scanning configuration. Just to outline some of the instrumental errors, σ_θ is always between 9% and 11%, c_1 between 9% and 13% and c_0 4% with exception to S4.

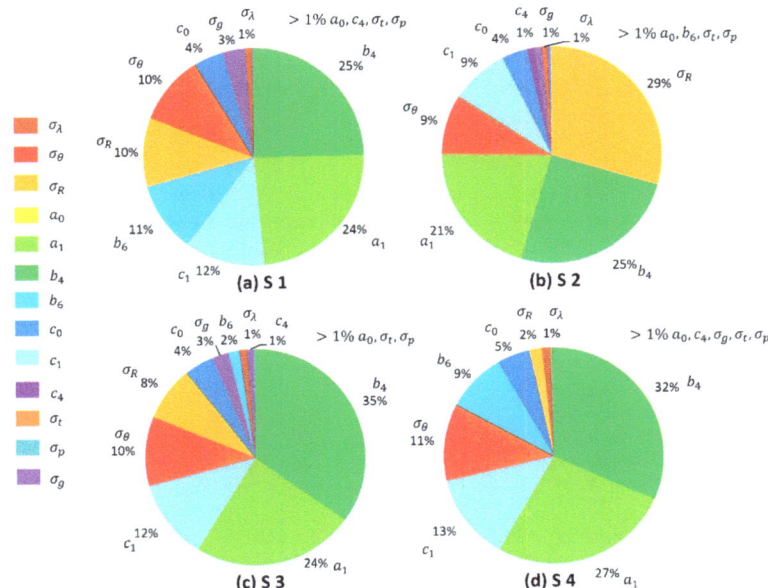

Figure 8. The contribution of variances for instrumental and atmospheric elementary errors in percent for (**a**) S1, (**b**) S2, (**c**) S3, (**d**) S4. For parameter names, see Sections 3.2 and 3.4.

The elementary errors for air temperature and air pressure at this level of variance (see Table 3) fall into the same category remaining under the quote of 1%, fact related to the small influencing coefficients in Equation (11). The contribution of the VGT reaches 3% of the error budget for S1 (Figure 8a) and S3 (Figure 8c). This is explainable by the fact that a large part of the observations are in the lower atmosphere layer, where the VGT instability is known to be higher (cf. [53]). A higher level of variance would generally lead to higher contributions to the error budget, but this is planned to be studied further for longer ranges and within different timespans.

5. Conclusions

Throughout this contribution, a method of defining a stochastic model for TLS observations was presented. At the time being, this model considers instrumental errors of long-range laser scanners and meteorological error sources.

The EEM is improved by directly introducing the existing spatial correlations into the synthetic VCM, without the need to compute them each time. This will confirm the EEM advantages when a second measurement epoch is available. It was also shown that some of the instrumental and atmospheric elementary errors can be neglected at the given level of variance, having a contribution of less than 1% to the complete error budget. Within this contribution, the line of previous EEM publications was continued with the introduction of atmospheric elementary errors treated as stochastic correlating and the introduction of a functional model for hybrid laser scanners. The newly achieved VCM plays an important role in surface estimations as already shown in [7]. Other possible applications that may benefit from this model can be encountered in landslide, glacier or rock cliff monitoring.

Resuming the newly discussed topics, it can be mentioned that:

- the functional model was adapted for the instrumental errors of the Riegl VZ-2000 scanner;
- a deterministic approach was used to consider the spatial distribution of the atmospheric errors;
- the atmospheric elementary errors were included in the EEM as stochastic correlating errors;

- the error of position was presented in relation to the scanner position (geometric configuration) and yielded average values between 9 mm and 37 mm for ranges of 88 m to 456 m;
- spatial correlations have been analyzed with respect to a vertical section in each case;
- the contribution of individual error sources outlined the fact that instrumental errors have the biggest impact on the error of position.

As regards the topics under research, the EEM for TLS measurements is still lacking object related elementary errors. This implies using existing studies for different materials scanned from different positions at different ranges and include these errors into the stochastic correlating group. The authors intend to use intensity values of the reflected laser beam, as introduced by Wujanz et al. [61]. Having done this would make the EEM a powerful tool for generating a TLS stochastic model that considers all important error sources. Another topic under research is the sensitivity analysis of the input parameters. After completing the stochastic model, attention will be accorded to the impact of each individual input parameter on the outcomes. This allows an estimation of the optimal scanning configuration.

Author Contributions: Methodology, G.K. and V.S.; software, G.K.; validation, V.S. and G.K.; formal analysis, G.K.; resources, G.K. and V.S.; data acquisition, G.K.; writing—original draft preparation, G.K.; writing—review and editing, V.S.; supervision, V.S.; project administration, V.S. All authors have read and agreed to the published version of the manuscript.

Funding: This research was funded by DFG (German Research Foundation), SCHW 838/7-3 within the project IMKAD II "Integrated space-time modeling based on correlated measurements for the determination of survey configurations and the description of deformation processes". The project is also associated with Cluster of Excellence Integrative Computational Design and Construction for Architecture (IntCDC) and supported by the DFG under Germany's Excellence Strategy—EXC 2120/1—390831618.

Acknowledgments: The authors also express their gratitude to Illwerke vkv AG, especially thanking Ralf Laufer for supporting the measurement campaign in 2019. Acknowledgment also goes to "Landesamt für Vermessung und Geoinformation" Land Vorarlberg for making LiDAR data available for the research area around the Kops lake.

Conflicts of Interest: The authors declare no conflict of interest.

References

1. Uren, J.; Price, B. *Surveying for Engineers*, 5th ed.; Palgrave Macmillan: New York, NY, USA, 2010.
2. Kuhlmann, H.; Schwieger, V.; Wieser, A.; Niemeier, W. Engineering Geodesy-Definition and Core Competencies. *J. Appl. Geod.* **2014**, *8*, 327–334. [CrossRef]
3. Ogundare, J.O. *Precision Surveying: The Principles and Geoamtics Practice*; John Wiley & Sons, Inc.: Hoboken, NY, USA, 2016; pp. 8–13.
4. Wunderlich, T.; Niemeier, W.; Wujanz, D.; Holst, C.; Neitzel, F.; Kuhlmann, H. Areal Deformation Analysis from TLS Point Clouds–The Challenge. *Allg. Vermess.* **2016**, *123*, 340–351.
5. Paffenholz, J.A.; Stenz, U.; Wujanz, D.; Neitzel, F.; Neumann, I. 3DPunktwolken-basiertes Monitoring von Infrastrukturbauwerken am Beispiel einer historischen Gewölbebrücke. In Proceedings of the TLS 2017 Seminar DVW-Schriftreihe, Band 88/2017, Fulda, Germany, 11–12 December 2017.
6. Kermarrec, G.; Alkhatib, H.; Neumann, I. On the Sensitivity of the Parameters of the Intensity-Based Stochastic Model for Terrestrial Laser Scanner. Case Study: B-Spline Approximation. *Sensors* **2018**, *18*, 2964. [CrossRef] [PubMed]
7. Harmening, C.; Neuner, H. A spatio-temporal deformation model for laser scanning point clouds. *J. Geod.* **2020**, *94*, 26. [CrossRef]
8. Mettenleiter, M.; Härtl, F.; Kresser, S.; Fröhlich, C. *Laserscanning—Phasenbasierte Lasermesstechnik für die Hochpräzise und Schnelle Dreidimensionale Umgebungserfassung*; Süddeutscher Verlag onpact GmbH: Munich, Germany, 2015; pp. 4–5.
9. Staiger, R. Terrestrial Laser Scanning Technology-Systems and Applications. In Proceedings of the 2nd FIG Regional Conference, Marrakech, Morocco, 2–5 December 2003.

10. Wieser, A.; Pfaffenholz, J.-A.; Neumann, I. Sensoren, Features und Physik-Zum aktuellen Stand der Entwicklung bei Laserscannern. In Proceedings of the TLS 2019 Seminar, DVW-Schriftreihe, Band 96/2019, Fulda, Germany, 2–3 December 2019.
11. Kuhlmann, H.; Holst, C. Flächenhafte Abtastung mit Laserscanning-Messtechnik, flächenhafte Modellierung und aktuelle Entwicklungen im Bereich des terrestrischen Laserscanning. In *Ingenieurgeodäsie-Handbuch der Geodäsie*; Schwarz, W., Ed.; Springer: Berlin, Germany, 2018; pp. 203–205.
12. Neuner, H.; Holst, C.; Kuhlmann, H. Overview on Current Modelling Strategies of Point Clouds for Deformation Analysis. *Allg. Vermess.* **2016**, *123*, 328–339.
13. Borovkov, K. *Elements of Stochastic Modelling*; World Scientific Publishing Co. Pte. Ltd.: Singapore, 2014; pp. 3–4.
14. Niemeier, W. *Ausgleichungsrechnung*, 2nd ed.; Walter de Gruyter: Berlin, Germany, 2008; pp. 31–33.
15. Matthias, H.J. Bedeutung und Konstruktion von Kovarianzen in der Messtechnik. Ph.D. Thesis, Institut für Geodäsie und Photogrammetrie an der Eidgenössischen Technischen Hochschule Zürich, Mitteilungen Nr. 41, Zürich, Switzerland, 1992.
16. Zhao, X.; Kermarrec, G.; Kargoll, B.; Alkhatib, H.; Neumann, I. Influence of the simplified stochastic model of TLS measurements on geometry-based deformation analysis. *J. Appl. Geod.* **2019**, *13*, 199–214. [CrossRef]
17. Kauker, S.; Schwieger, V. A synthetic covariance matrix for monitoring by terrestrial laser scanning. *J. Appl. Geod.* **2017**, *11*, 77–87. [CrossRef]
18. Schweitzer, J.; Schwieger, V. Modeling and Propagation of Quality Parameters in Engineering Geodesy Processes in Civil Engineering. In Proceedings of the 1st International Workshop on the Quality of Geodetic Observation and Monitoring Systems, Munich, Germany, 13–15 April 2011; Kutterer, H., Seitz, F., Alkhatib, H., Schmidt, M., Eds.; Springer: Cham, Switzerland, 2015.
19. *ISO/IEC Guide 98-3:2008: Uncertainty of Measurement—Part 3: Guide to the Expression of Uncertainty in Measurement (GUM:1995)*; International Organization for Standardization: Genève, Switzerland, 2008.
20. Metropolis, N.; Ulam, S. The Monte Carlo Method. *J. Am. Stat. Assoc.* **1949**, *44*, 335–341. [CrossRef] [PubMed]
21. Pelzer, H. Grundlagen der Mathematischen Statistik und der Ausgleichungsrechung. In *Geodätische Netze in Landes- und Ingenieurvermessung*; Pelzer, H., Ed.; Konrad Wittwer: Stuttgart, Germany, 1985.
22. Koch, K.R. Determining uncertainties of correlated measurements by Monte Carlo simulations applied to laserscanning. *J. Appl. Geod.* **2008**, *2*, 139–147. [CrossRef]
23. Aichinger, J.; Schwieger, V. Influence of scanning parameters on the estimation accuracy of control points of B-spline surfaces. *J. Appl. Geod.* **2018**, *12*, 157–167. [CrossRef]
24. Förstner, W. Ein Verfahren zur Schätzung von Varianz-und Kovarianzkomponenten. *Allg. Vermess.* **1979**, *86*, 446–453.
25. Koch, K.R. *Parameterschätzung Und Hypothesentests in Linearen Modellen*, 1st ed.; Ed. Dümmler: Bonn, Germany, 1980; pp. 245–259.
26. Teunissen, P.J.G.; Amiri-Simkooei, A.R. Least-squares variance component estimation. *J. Geod.* **2008**, *82*, 65–82. [CrossRef]
27. Niemeier, W.; Hollman, R. *Haputkomponenten- und Sensitivitätsanalyse Geodätischer Netze aufgezeigt am Überwachungsnetz Varna*; WAdFV Hanover, Nr. 133; Hanover, Germany, 1984; pp. 61–72.
28. Saltelli, A.; Chan, K.; Scott, E.M. (Eds.) *Sensitivity Analysis*, 1st ed.; John Wiley and Sons: New York, NY, USA, 2000.
29. Hagen, G. *Gründzuge der Wahrscheinlichkeits-Rechnung*; Ed. Dümmler: Berlin, Germany, 1837.
30. Bessel, F.W. Untersuchung über die Wahrscheinlichkeit der Beobachtungsfehler. *Astron. Nachr.* **1837**, *15*, 369–404.
31. Schwieger, V. *Ein Elementarfehlermodell für GPS Überwachungsmessungen*; Schriftenreihe der Fachrichtung Vermessungswesen der Universität Hannover: Hanover, Germany, 1999; Volume 231.
32. Augath, W. Lagenetze, Geodätische Netze in Landes-und Ingenieurvermessung. In *Geodätische Netze in Landes-und Ingenieurvermessung*; Pelzer, H., Ed.; Konrad Wittwer: Stuttgart, Germany, 1985.
33. Schwieger, V. Determination of Synthetic Covariance Matrices–An Application to GPS Monitoring Measurements. In Proceedings of the 15th European Signal Processing Conference EUSIPCO 2007, Poznan, Poland, 3–7 September 2007.
34. Rüeger, J.M. *Electronic Distance Measurement: An Introduction*, 3rd totally revised ed.; Springer: Berlin, Germany, 1990.

35. Jutzi, B. Analyse der zeitliche Signalform von Rückgestreuten Laserpulse. Ph.D. Thesis-DGK, Reihe C, Heft Nr. 611, Technical University of Munich, Munich, Germany, 2007.
36. Soudarissanane, S.S. The Geometry of Terrestrial Laser Scanning-identification of Errors, Modeling and Mitigation of Scanning Geometry. Ph.D. Thesis, Technical Universit of Delf, Delft, The Netherlands, 2016.
37. Chow, J.C.K.; Lichti, D.D.; Glennie, C.; Hartzell, P. Improvements to and Comparison of Static Terrestrial LiDAR Self-Calibration Methods. *Sensors* **2013**, *13*, 7224–7249. [CrossRef] [PubMed]
38. Riegl Laser Measurement Systems GmbH, Horn, Austria. Datasheet of Riegl VZ-2000. Available online: http://www.riegl.com (accessed on 20 January 2020).
39. Rabinovich, S.G. *Evaluating Measurement Accuracy—A Practical Approach*, 3rd ed.; Springer International Publishing AG: Cham, Switzerland, 2017.
40. Wunderlich, T.H.; Wasmeier, P.; Ohlmann-Lauber, J.; Schäfer, T.H.; Reidl, F. Objective Specifications of Terrestrial Laserscanners—A Contribution of the Geodetic Laboratory at the Technische Universität München. *Blaue Reihe des Lehrstuhls für Geodäsie* **2013**, *21*, 8–10.
41. Lichti, D. Error modeling, calibration and analysis of an AM-CW terrestrial laser scanner system. *ISPRS J. Photogramm. Remote Sens.* **2007**, *66*, 307–324. [CrossRef]
42. Lichti, D. Terrestrial laser scanner self-calibration: Correlation sources and their mitigation. *ISPRS J. Photogramm. Remote Sens.* **2010**, *65*, 93–102. [CrossRef]
43. Schneider, D. Calibration of a Riegl LMS-Z420i based on a Multi-Station Adjustment and a Geometric Model with Additional Parameters. In Proceedings of the Laser Scanning 2009, Paris, France, 1–2 September 2009; Bretar, F., Pierrot-Deseilligny, M., Vosselman, G., Eds.; IAPRS: Paris, France, 2009; Vol. XXXVIII, Part 3/W8.
44. Fröhlich, C. Aktive Erzeugung korresnpondierender Tiefen-und Reflektivitätsbilder und ihre Nutzung zur Umgebungserfassung. Doctoral Thesis, Leibniz Universität Hannover, Munich, Germany, 1996.
45. Riegl Laser Measurement Systems GmbH, Horn, Austria RiSCAN Pro Software Help Documentation. 2015. Available online: http://www.riegl.com/products/software-packages/riscan-pro/ (accessed on 20 January 2020).
46. International Association of Geodesy. IAG Resolutions adopted at the XXIIth General Assembly in Birmingham 1999. Available online: https://iag.dgfi.tum.de/fileadmin/IAG-docs/IAG_Resolutions_1999.pdf (accessed on 19 January 2020).
47. Reshetyuk, Y. Self-Calibration and Direct Georeferencing in Terrestrial Laser Scanning. Ph.D. Thesis, Universitetsservice, US AB, Stockholm, Sweden, 2009.
48. Friedli, E.; Presl, R.; Wieser, A. Influence of atmospheric refraction on terrestrial laser scanning at long range. In Proceedings of the 4th Joint International Symposium on Deformation Monitoring (JISDM), Athens, Greece, 15–17 May 2019.
49. Joeckel, R.; Stober, M.; Huep, W. *Elektronische Entferungs- und Richtungsmessung und ihre Integration in aktuelle Positionierungsverfahren*; Wichmann Verlag: Heidelberg, Germany, 2008; pp. 221–225.
50. Kahmen, H. *Angewandte Geodäsie: Vermessungskunde*, 20th ed.; Walter de Gruyter: Berlin, Germany, 2006; pp. 459–460.
51. Brunner, F.K. *Geodetic Refraction: Effects of Electromagnetic Wave Propagation through the Atmosphere*; Springer: Berlin, Germany, 1984; pp. 2–3.
52. Brocks, K. *Vertikaler Temperaturgradient und terrestrische Refraktion, insbesondere im Hochgebirge*; Band III, Heft 4; Publications Meteorological Institute University of Berlin: Berlin, Germany, 1939.
53. Geiger, R.; Aron, R.H.; Todhunter, P. *The Climate near the Ground*, 6th ed.; Rowman & Littlefield Publishers, Inc.: Oxford, UK, 2003.
54. Hirt, C.; Guillaume, S.; Wisbar, A.; Bürki, B.; Sternberg, H. Monitoring of the refraction coefficient in the lower atmosphere using a controlled setup of simultaneous reciprocal vertical angle measurements. *J. Geophys. Res.* **2010**, *115*, D21102. [CrossRef]
55. Hennes, M. Das Nivelliersystem-Feldprüfverfahren nach ISO 17123-2 im Kontext refraktiver Störeinflüsse. *Allg. Vermess.* **2006**, *3*, 85–94.
56. Deutscher Wetterdienst. URL. Available online: https://www.dwd.de/DE/service/lexikon/lexikon_node.html (accessed on 20 February 2020).
57. Beraldin, J.A.; Blais, F. Laser Scanning Technology. In *Airborne and Terrestrial Laser Scanning*; Vosselman, G., Maas, H.-G., Eds.; Whittles Publishing: Dunbeath, UK, 2010; pp. 14–15.
58. Illwerke vkw, AG. Available online: https://www.illwerkevkw.at/kopssee.htm (accessed on 22 January 2020).

59. Kerekes, G.; Schwieger, V. Determining Variance-covariance Matrices for Terrestrial Laser Scans: A Case Study of the Arch Dam Kops. In in Proceedings of the INGEO & SIG 2020, Dubrovnik, Croatia, 1–4 April 2020. Before print.
60. GHM Messtechnik GmbH. Product specifications GTD 1100. 2020; Available online: https://www.greisinger.de/files/upload/de/produkte/kat/k19_095_DE_oP.pdf (accessed on 20 January 2020).
61. Wujanz, D.; Burger, M.; Mettenleiter, M.; Neitzel, F. An intensity-based stochastic model for terrestrial laser scanners. *ISPRS J. Photogramm. Remote Sens.* **2017**, *125*, 146–155. [CrossRef]

© 2020 by the authors. Licensee MDPI, Basel, Switzerland. This article is an open access article distributed under the terms and conditions of the Creative Commons Attribution (CC BY) license (http://creativecommons.org/licenses/by/4.0/).

Article

On Estimating the Hurst Parameter from Least-Squares Residuals. Case Study: Correlated Terrestrial Laser Scanner Range Noise

Gaël Kermarrec

Geodetic Institute, Leibniz Universität Hannover, Nienburger Str. 1, 30167 Hannover, Germany; kermarrec@gih.uni-hannover.de; Tel.: +49-511-7621-4736

Received: 27 February 2020; Accepted: 26 April 2020; Published: 29 April 2020

Abstract: Many signals appear fractal and have self-similarity over a large range of their power spectral densities. They can be described by so-called Hermite processes, among which the first order one is called fractional Brownian motion (fBm), and has a wide range of applications. The fractional Gaussian noise (fGn) series is the successive differences between elements of a fBm series; they are stationary and completely characterized by two parameters: the variance, and the Hurst coefficient (H). From physical considerations, the fGn could be used to model the noise of observations coming from sensors working with, e.g., phase differences: due to the high recording rate, temporal correlations are expected to have long range dependency (LRD), decaying hyperbolically rather than exponentially. For the rigorous testing of deformations detected with terrestrial laser scanners (TLS), the correct determination of the correlation structure of the observations is mandatory. In this study, we show that the residuals from surface approximations with regression B-splines from simulated TLS data allow the estimation of the Hurst parameter of a known correlated input noise. We derive a simple procedure to filter the residuals in the presence of additional white noise or low frequencies. Our methodology can be applied to any kind of residuals, where the presence of additional noise and/or biases due to short samples or inaccurate functional modeling make the estimation of the Hurst coefficient with usual methods, such as maximum likelihood estimators, imprecise. We demonstrate the feasibility of our proposal with real observations from a white plate scanned by a TLS.

Keywords: terrestrial laser scanner; stochastic model; B-spline approximation; Hurst exponent; fractional Gaussian noise; generalized Hurst estimator

1. Introduction

Terrestrial laser scanners (TLS) capture a large amount of 3D points rapidly, with high precision and spatial resolution [1]. These scanners are used for applications as diverse as modeling architectural and engineering structures, and high-resolution mapping of terrain, vegetation, and other landscape features. The recorded point clouds can be processed and analyzed with dedicated software. In engineering geodesy, this processing allows for the computation of deformation magnitudes. Unfortunately, these latter are negatively affected when noisy and scattered point clouds (PC) are used. Additionally, no rigorous statistical test for deformation can be performed with the raw PC [2].

These drawbacks can be circumvented by approximating the PC with mathematical surfaces [3]. Besides norms such as L1 or L∞ [4], a widely used criterion is the sum of squares of the orthogonal distances from the data points to the parametric surface. Exemplarily, regression B-spline enjoys special attention to approximate point clouds from TLS: B-splines basis functions have a closed form expression, are polynomial, and, thus, particularly easy to compute (see [5] for one of the first articles related to that topic in geodesy). The setup of specific statistical tests with confidence intervals is based on the estimated parameters, or on the approximated surface points. Exemplarily, the congruence test can be used to test for deformation ([2]) and is known to be the most powerful test in Gauss–Markov models with normally distributed random deviations and a correctly specified stochastic model. The setting of a realistic variance covariance matrix (VCM) of the raw observations of the TLS is done prior to this test [6].

As for every sensor recording millions of points in a few minutes, the measurements of TLS are expected to be temporally correlated. Physically, the range or distance measurements are phase differences, so that a power law spectral density of the correlated range noise is hardly plausible ([7], [8]). This correlation structure was empirically proven in a few recent real case analyses, see, e.g., [9] or [10]. The authors approximated simple scanned objects with a Gauss–Helmert model [11], assuming pre-defined geometric primitives such as circle, ellipsoid, and plane. The correlation parameters were estimated by fitting the residuals of the approximation with an autoregressive function of the first order (AR(1)). This methodology has, however, drawbacks:

(i) the exponential covariance function restricts the description of the correlation structure to short range dependency and may not be a physically adequate modelization,
(ii) empirically the covariance can be problematic. Two autocovariance functions sharing a common principal irregular term won't yield asymptotically the same best linear predictor (see [12], chapter 3) for some examples. Using the popular Gaussian function can lead to overoptimistic predictions. Additionally, antipersistent data are difficult to distinguish from uncorrelated data and the correlations can be mistaken for noise fluctuations around zero,
(iii) the methodology could be made more general: it is based on a calibrated object scanned in a controlled environment.

In this contribution, we propose to address these drawbacks and to derive a general methodology to assess the correlation structure of the TLS range measurements. We will base ourselves on the physical expectation that the TLS range noise should have a long-range dependency (LRD) and heavy tailed distribution. Our proposal to extract the correlation structure is applicable to every kind of object, without being restricted to predefined objects or calibration scenarios. It is extendable to other kinds of observations, such as residuals from a geodetic coordinate time series [13].

We choose to model the noise of TLS range with a stationary LRD noise: the fractional Gaussian noise (fGn), which is entirely defined by the Hurst parameter (abbreviated by H) and the variance. It has the main advantage that the autocorrelation function can be easily estimated without computation burden [14]. Fractal time series or signals such as the fGn have been found in many domains, including biology [15], medicine (EEG [16,17]), finance (stock market analysis [18]), geology, and traffic analysis [19]. Various statistical techniques have been proposed to estimate H and each has shortcomings and advantages: they may perform better in the presence of noise, for short samples, or for H close to a given value, may have slow convergence, etc. (see, e.g., [20–24]). There exist three families of estimation methods: the time domain (e.g., Rescaled Range R/S estimator [25], the detrended fluctuation analysis method [26]), the frequency domain (periodogram [27], the Whittle estimator [28]), and the wavelet space [29], which was shown to provide an unbiased, efficient, and robust estimator.

We will use the residuals of the B-splines approximation of the TLS point clouds to assess the correlation structure of TLS range. We conjecture that although (i) additional white noise and (ii) possible model misspecification could introduce additional frequencies in the residuals, these latter still contain enough information to estimate H, provided that an adequate filtering is performed. Besides the simulated observations from a TLS, we will evaluate our methodology and compare the performance of three different estimators for H using real observations from a white plane.

We firstly disregard the correlations of the polar angles; a similar methodology as presented in this present study could be used to that aim.

The remainder of this paper is structured as follows: the first section provides a brief summary of the mathematical concepts of least-squares and stochastic modeling. The second section introduces the concepts of fGn, Hurst exponent, and filtering. The third section describes the results of simulations for two specific cases: a plane and a Gaussian curve. We conclude with a real case study and some recommendations.

2. Mathematical Background of Surface Fitting

2.1. Functional Model

Free-form curves and surface fittings are flexible tools to approximate PC without being restricted by the use of geometric primitives, such as circle, planes, or cylinders. Possible applications of surface approximation include the testing of deformation [30] or the reduction of a huge amount of points to a simpler form.

In this contribution, we make use of B-spline surfaces. Their properties and advantages over other functions, such as control and flexibility, are exemplarily described in [31]. Readers interested in more details on how spline fitting works should refer, e.g., to [32,33], and more specifically for geodetic applications to [5] or [31]. Such surfaces satisfy the strong convex hull property and have a fine and local shape control so that they were shown to be adequate for approximating noisy and scattered PC (see, e.g., [34]). For the sake of shortness, we shortly introduce the main concept, focusing on least-squares (LS) approach to determine the model parameters, called control points (CP).

We start with n_{obs} polar observations from a TLS expressed in vector form l_{POLAR} of size ($3n_{obs}$). The observations are made of two angles, HA and VA, and one range ρ to which a VCM Σ_{ll_POLAR} is associated. This matrix describes the variance and possible correlation between the observations [11] and is focus of our contribution.

2.1.1. First Step: From Polar to Cartesian

The first step of the approximation is the transformation of the PC coordinates vector from polar l_{POLAR} into Cartesian l_{CART}.

The VCM has to be transformed by the error propagation law. The VCM Σ_{ll_CART} reads:

$$\Sigma_{ll\_CART} = \mathbf{F} \Sigma_{ll\_POLAR} \mathbf{F}^T \qquad (1)$$

The matrix $\mathbf{F}$ contains the derivatives of the point coordinates with respect to the range and angles and is given for one point i by:

2.1.2. Second Step: The Approximation

The Cartesian PC can be approximated mathematically by means of a linear combination of basis functions, such as B-splines. In its parametric formulation, the B-spline surface $\mathbf{s}(t,f)$ is a tensor product surface and can be expressed as

$$\mathbf{s}(t,f) = \sum_{i=0}^{n} \sum_{j=0}^{m} N_{i,p_b}(t) N_{j,q_b}(f) \mathbf{p}_{i,j}, \qquad (2)$$

where $(t,f) \in [0,1] \times [0,1]$ are the parameters in the two directions so that a B-spline surface maps the unit square to a rectangular surface patch. The basis function $N_{i,p}$ and $N_{j,q}$ are composite curves of degree p and q polynomials, respectively, with joining points at knots in the interval $[u_i, u_{i+p+1}]$ and $[v_j, v_{j+q+1}]$. They can be evaluated by means of a recurrence relationship [32]. To summarize, the surface is defined by:

- a set of n + 1 CP in the direction t and m + 1 CP in the direction f,
- a knot vector of h + 1 knots in the t-direction, $\mathbf{U} = [u_0, \ldots, u_h]$,
- a knot vector of k + 1 knots in the f-direction $\mathbf{V} = [v_0, \ldots, v_k]$,

- the degree p_b of the basis functions in the t-direction, and the degree q_b in the f-direction.

In this contribution, we will take a degree of $p_b = q_b = 3$ for the B-splines functions (cubic B-splines). We solve the determination of an optimal knot vector using the knot placement technique as described in [33]. The Cartesian point cloud is parametrized in advance with the equidistant parametrization, justified by the simple structure of the objects under consideration in this contribution.

2.1.3. Third Step: LS Solution

Approximating a PC with a B-spline surface is finding the coordinates of the CP so that the distance of the data points to the approximated surface is minimized. This step can be performed by solving the LS problem, for which the minimum in the LS sense of the zero-mean error term $\mathbf{v}$ is searched: $\min_{p \in \mathbb{R}^3} \|\mathbf{v} = \mathbf{A}\mathbf{p} - \mathbf{l}_{CART}\|^2_{\Sigma_{ll\_CART}}$.

$\mathbf{p}$ is the matrix of CP to be estimated and is of size $(3(n+1)(m+1))$, $\mathbf{A}$ –$(3n_{obs}, 3(n+1)(m+1))$– is called the design or mass matrix. It contains the evaluation of the B-spline functions at the parameters. Interested readers should refer to [35] for the description of the design matrix.

The estimated coordinates of the control points are expressed by the unbiased generalized LS estimator (GLSE [11]):

$$\hat{\mathbf{p}}_{GLSE} = \left(\mathbf{A}^T \mathbf{\Sigma}^{-1}_{ll\_CART} \mathbf{A}\right)^{-1} \mathbf{A}^T \mathbf{\Sigma}^{-1}_{ll\_CART} \mathbf{l}_{CART} \tag{3}$$

If the VCM Σ_{ll_CART} is the identity matrix (equal variance for all coordinates), the ordinary LS estimator (OLSE) is obtained: $\hat{\mathbf{p}}_{OLSE} = \left(\mathbf{A}^T \mathbf{A}\right)^{-1} \mathbf{A}^T \mathbf{l}_{CART}$.

We further note that the LS estimator is unbiased $E(\hat{\mathbf{p}}_{GLSE}) = E(\hat{\mathbf{p}}_{OLSE}) = \mathbf{p}$ so that $\hat{\mathbf{p}}$ can be computed either with the OLSE and GLSE solution; thanks to the unbiasedness of the LS estimator, the expectation E of the estimated coordinates of the CP are not affected by the choice of Σ_{ll_CART}. However, the OLSE is not the most efficient within the class of linear unbiased estimators anymore when Σ_{ll_CART} deviates from the true (and unknown) VCM. Consecutively, hypothesis tests such as the global test, outlier tests, or congruence tests become invalid [36]. It is one of the main reasons why assessing the correlation structure of the raw measurements is an actual research topic.

The number of control points has an impact on the LS solution on the fitted surface. It can be either fixed a priori or iteratively adjusted in the context of model selection [37]. As the impact of model misspecifications is interesting for our purpose, we will make use of the first strategy.

2.2. The Residuals of the LS Surface Approximation

2.2.1. The Cartesian Residuals

We call $\hat{\mathbf{v}}_{CART} = \mathbf{A}\hat{\mathbf{p}} - \mathbf{l}_{CART}$ the residuals of the LS adjustment. $\hat{\mathbf{l}}_{CART} = \mathbf{H}\hat{\mathbf{p}}$ are the adjusted observations with $\mathbf{H}$ being the Hat matrix, $\mathbf{H} = \mathbf{A}\left(\mathbf{A}^T \mathbf{\Sigma}^{-1}_{ll\_CART} \mathbf{A}\right)^{-1} \mathbf{A}^T \mathbf{\Sigma}^{-1}_{ll\_CART}$. The VCM of the adjusted residuals reads:

$$\Sigma_{\hat{\mathbf{v}}_{CART} \hat{\mathbf{v}}_{CART}} = (\mathbf{I} - \mathbf{H})^T \mathbf{\Sigma}^{-1}_{ll\_CART} (\mathbf{I} - \mathbf{H}) \tag{4}$$

with $\mathbf{I}$ being the identity matrix.

We further defined the a posteriori variance factor as

$$\hat{\sigma}_0^2 = \frac{\hat{\mathbf{v}}_{CART}^T \mathbf{\Sigma}^{-1}_{ll\_CART} \hat{\mathbf{v}}_{CART}}{n_{obs} - 3(n+1)(m+1)}, \tag{5}$$

This factor can be used to judge the goodness of fit of the LS adjustment by means of a global test [36]. The a priori VCM of the estimates is given by $\Sigma_{\hat{\mathbf{p}}\hat{\mathbf{p}}} = \left(\mathbf{A}^T \mathbf{\Sigma}^{-1}_{ll\_CART} \mathbf{A}\right)^{-1}$.

2.2.2. The Polar Residuals

In this contribution, we propose to extract the correlation structure of the TLS range from the LS residuals of the B-spline approximation. We answer the drawback raised in the introduction by being independent of calibrated objects, i.e., our methodology should be applicable in every environment.

As mentioned in Section 2.2.1, the LS adjustment gives access to the Cartesian residuals. To assess the noise of the raw TLS observations (range), we transform the Cartesian residuals into polar $\hat{\mathbf{v}}_{POLAR} = \begin{bmatrix} \hat{\mathbf{v}}_{HA}, \hat{\mathbf{v}}_{VA}, \hat{\mathbf{v}}_r \end{bmatrix}$. These latter have a VCM $\Sigma_{\hat{\mathbf{v}}_{POLAR}\hat{\mathbf{v}}_{POLAR}}$ obtained similarly to Equation (1), using the error propagation law "backwards". This matrix depends on the original matrix Σ_{ll_POLAR}. Our assumption is that we can derive the correlation structure of the original observations from $\hat{\mathbf{v}}_{POLAR}$. We will further focus on $\hat{\mathbf{v}}_r$ and conjecture that this final vector still contains enough information to give us access to the approximate correlation structure of Σ_{ll_r}, defined as the VCM of the range observations.

3. Noise Description of TLS Range Observations: Variance and Correlation

3.1. Variance

Raw observations from TLS are the three polar coordinates of the recorded points. They are made of a range and two angles in the vertical and horizontal direction. These observations are known to have different noise properties ([38–40]): the noise of angles is widely assumed to be Gaussian with a variance taken from manufacturer datasheets. The noise of the range measurements has a slightly different structure. Its variance can be considered as a constant; the manufacturer datasheets provide different values depending on, e.g., the approximated distance to the scanned object and/or to the properties of the surface (roughness, color). Alternatively, the variance can be modeled as following a point-wise power law intensity model [41,42].

In this contribution, we simulate different point clouds with a noise variance close to what is expected in a real case experiment: $\begin{cases} \sigma_{HA} = \sigma_{VA} = 0.0001 \\ \sigma_\rho = 0.005 m \end{cases}$, where $\sigma_{HA} = \sigma_{VA}$ are the standard deviations for the HA and VA and σ_ρ for the range, respectively. We intentionally chose a case where range and angle have different variances to simulate a more general scenario.

3.2. Correlation Structure for Range Measurements

In a first approximation, the range measurements are considered to be uncorrelated, i.e., one observation recorded at time t is not dependent on the observation recorded at $t + \tau$, τ being the interval between two measurements. τ is also called time lag; it is related to the scanning rate of the observations and depends on the setting. Exemplarily, the resolution for a TLS Z+F 5016 can be varied from preview to ultrahigh up to extremely high, and low to high quality: these choices impact the scanning rate, and thus the scanning time of an object. The assumption of uncorrelatedness is overoptimistic: range measurements are based on phase differences, which are inherently influenced by, e.g., the propagation of the signal through a random media, but also by the point spacing on the surface.

In this contribution, the correlation structure of the range will be modeled as a fractional Gaussian noise (fGn). This assumption is justified by the physically based expectation of the author that the range noise is stationary and that its power spectrum will follow a power law [7]. The validation of this model with real data is shortly shown in Section 4.5. More extensive works using TLS observations will be performed in a next step based on the proposed methodology.

fGn has the beneficial properties that it is characterized by its variance and a single parameter called the Hurst exponent H. We will here shortly introduce the concept of fGn; interested readers can refer to [14] for more information.

3.2.1. What is a fGn?

To define a fGn, the understanding of the LRD concept is mandatory. This property of a process is linked with the slow decay of the autocorrelation ρ to zero so it is a non-summable function, i.e., if the average value of its partial sums does not converge, see [43]. More precisely,

$$\rho(\tau) \approx c\tau^{-\delta}, \tag{6}$$

with τ the time lag, c a positive constant, and $0 < \delta < 1$. As τ increases, the dependence between the observations stays strong, which implies a fat tailed autocorrelation function. Exemplarily, for a stationary process $\delta = 0.3$, the autocorrelation for lag 100 will stay at 0.15, whereas for a Markovian process, the autocorrelation would be practically zero for lags 10 times less. This important property is the reason why such processes are said to have a "long-term memory" and it is one of the major reasons why we wish to model the correlation structure of TLS range observations with such a process. Intuitively, the high rate of measurements induces a long dependency between the observations: the autocorrelation may decay quickly at the origin—e.g., between the first and the second observation—but stays for a long dependency much higher than 0. The autocorrelation will be similar between the first observation and the 100th or the 200th.

For a stationary process, the LRD can be related to a parameter called the Hurst exponent H, defined as a measure self-similarity. A stochastic process $X_H(t)$ is self-similar if $X_H(t)$ has the same distribution as $\lambda^{-H} X_H(\lambda t)$, where λ is a scale parameter. Concretely, the process will appear statistically identical under rescaling of the time axis by a given factor and $X_H(t) \infty \lambda^H$; it lacks any characteristic time scale. This characteristic allows interpretation of H as a measure of the strength of dependence between the time points, or more loosely, how much space the signal "fills in".

A self-similar process with stationary increment $X_H(t+1) - X_H(t)$ has an autocorrelation $C_H(\tau)$ given by

$$C_H(\tau) = \frac{1}{2}\left(|\tau+1|^{2H} - 2|\tau|^{2H} + |\tau-1|^{2H}\right), \tag{7}$$

so that for $\tau \to \infty$, $C(\tau) \to H(2H-1)\tau^{2H-2}$, meaning that the process has a long-range dependency, see Equation (6) and [44].

From these definitions, one can define the Hermite process of first order called the fractional Brownian Motion (fBm, [14]) as a generalization of a Brownian motion for which $H = \frac{1}{2}$. It is a non-stationary process with stationary increments and possesses the long-term memory, also called persistency or positive correlations when $H > \frac{1}{2}$. When $H < \frac{1}{2}$, the process has short term memory, or similarly anti-persistency or negative correlations; the autocorrelation decays fast enough so that their sums converge to a finite value. Both processes are described by a fractal dimension D, which is related to the Hurst exponent by $D = 2 - H$ for a fBm [44].

Successive increments ς_H of a fBm are called fGn:

$$\varsigma_H(t) = X_H(t+1) - X_H(t), \tag{8}$$

A fGn is, thus, a zero mean stationary process, defined as the stationary increment of fBm.

The fGn is fully characterized by the Hurst exponent and the variance $\sigma^2_{\varsigma_H}$. The corresponding distribution is completely specified by its autocovariance function given by Equation (7).

H can be related to the power-law spectrum $P(f) \infty \frac{1}{f^\beta}$, with f, β being the frequency and the power law of the process, respectively. Exemplarily $\beta = 0$ corresponds to a white noise, $\beta = 1$ is a pink noise and $\beta = 2$ is the Brownian noise. For a fBm, H is related to β by $H = \frac{\beta-1}{2}$, with $1 < \beta < 3$ and for a fGn, $H = \frac{\beta+1}{2}$ with $-1 < \beta < 1$ [45]. Using real observations, it is important to check if the noise is fGn or fBm: using the Matlab built-in function to estimate the Hurst exponent can lead to a misinterpretation of the results when not accounted for.

The difference between a fBm and a fGn can be visualized in Figure 1, where fBm (Figure 1, right top) and fGn (Figure 1, left top) versus time with different H are simulated. They are given with their corresponding power spectral densities (PSD), which decay linearly in a logplot (Figure 1, right bottom). We visualize the aforementioned "fills in" property of the process, i.e., small H"fills in" significantly more space than $H = 0.9$, which is related with a higher fractal dimension D.

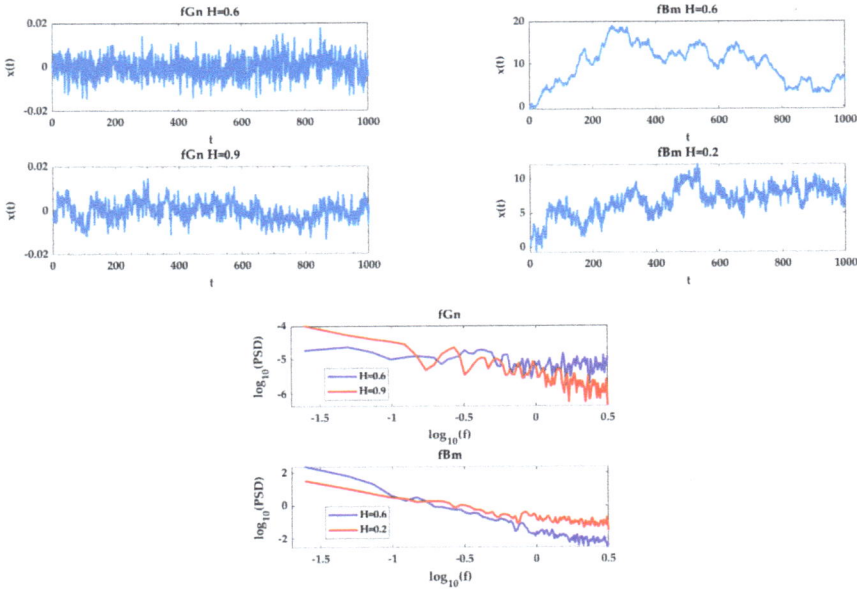

Figure 1. (**right,top**): Two realizations of a fGn with H = 0.6 (**top**) and H = 0.2 (**bottom**). (**left,top**): Two realizations of a fBm with H = 0.6 (**top**) and H = 0.9 (**bottom**). (**right,bottom**): The four corresponding PSD are plotted. Time series of 1000 observations were generated.

3.2.2. Generation of fGn

In this study, we are focusing on fractal stationary noise, i.e., fGn. From the previous section, and per the definition of the fGn, it can be generated by differentiating a fBm. Matlab (2018) provides a function called wfbm, which returns a fBm for a given Hurst parameter. This function uses the wavelet method from [46], which may bias the estimation of the Hurst parameter towards the wavelet method (described in the following section).

Alternatively, we propose to use the function called ffGn, a freely available function in the Matlab file exchange section. The ffGn function has the main advantage of being based on the circulant embedding method for persistent noise, resulting in a reproduction of its exact autocovariance [47]. To test the function, we assessed the standard deviation with which the Hurst parameter can be reproduced. We generated 10,000 realizations of short time series of lengths (1) 400 and (2) 1000. Focusing in this contribution on persistent fGn, the Hurst parameter was varied in the range of [1/2-1] by steps of 0.05. H was estimated using the three methods presented in the following sections. The standard deviation was found for all three methods to be between 0.01 for case (1) and 0.005 for case (2), highlighting the good performance and stability of the chosen function for noise generation.

3.2.3. How to Estimate the Hurst Parameter

Many methods have been proposed to estimate the Hurst exponent. They can be classified in three families: estimation in the time domain, frequency domain and wavelet domain. Intuitively, whereas the first family investigates the power-law relationship between a specific statistical property of the time-series and a time aggregation of it, the two latter examine if the spectrum or energy of the time-series follows power-law behavior. Inside these three families, different estimators have been proposed and constantly improved (see [48] and the references inside). We do not aim to review them all, which is exemplarily done in [49].

In this contribution, we focus on three estimators, which belong to each class: the generalized Hurst estimator, the Whittle estimator, and the wavelet estimator. The Whittle estimator was chosen for its capability to perform well when the number of observations is reduced; this case can occur when the TLS scanned lines are short, due to, e.g., the measurement configuration and the scanned object. The selection of the generalized Hurst estimator is justified by the fact that the noise of the angle or functional misspecifications may affect the power spectral density of the B-spline residuals at low frequency; this estimator focuses on the middle-high frequency part of the PSD and can be adapted with additional parameters. The wavelet estimator is said to be the less biased Hurst exponent estimator provided that a huge amount of observations is available (asymptotic behavior).

These challenges of estimating the correlation structure accurately are similar to the ones of the geodetic coordinate time series analysis (see [13] for further references on that topic). The chosen estimators have to account for these specificities. In the following, we will shortly review the three methods under consideration. A good understanding of their properties is mandatory to derive a meaningful methodology to extract an unknown H from B-spline residuals.

Generalized Hurst Exponent (GHE)

The generalized Hurst exponent was introduced in [50] and used for finance market analysis in [51] and [52]. It is a generalization of the approach proposed by [25].

The generalized Hurst exponent measures the LRD in the time domain. It is evaluated by using the qth-order moments of the distribution of increments: $K_q(t) = \frac{\langle |X_H(t+\tau) - X_H(t)|^q \rangle}{\langle |X_H(t)|^q \rangle}$.

τ is varied between 1 and $\tau_{\max}$ (usually taken to 20). Acting on $\tau_{\max}$ allows accounting for the specificity of the observations—for example, to force the estimator to focus on high frequencies—as described in [24]. $\langle \cdot \rangle$ stays for the average operator. H_q is related to $K_q(t)$: $K_q(\tau) \propto \tau^{qH_q}$, which allows the computation of

$$H_q \sim \frac{\log(K_q(\tau))}{q \log(\tau)}, \qquad (9)$$

as an average over a set of values corresponding to different τ. If H_q is not constant by varying q, the process is called multifractal, whereas $H_q = H$ characterizes an monofractal process [53]. In this contribution, and because we are not interested in the behavior of financial time series to predict the evolution of specific markets, we only estimate H_1, which describes the scaling behavior of the absolute values of the increments. H_1 reaches the value $\frac{1}{2}$ for a Gaussian noise. H_2 would correspond to an estimation in the frequency domain.

Whittle Likelihood Estimator (WhiE) Method

The Maximum Likelihood Estimator (MLE) is not a graphical method but is a purely numerical one. Thus, more than just the asymptotic self-similarity is assumed [53]; the MLE requires at least an assumption about the form of the LRD (such as a noise coming from fBm or Autoregressive integrated moving average ARIMA). If this assumption holds, it is often considered to be the best obtainable estimator; the estimates are asymptotically unbiased, and the estimator is asymptotically efficient and fast to compute. Unfortunately, MLE performs poorly if the assumption is incorrect or for short samples [54]. Exact maximum likelihood inference can be performed for Gaussian data ([55]) by evaluation the log-likelihood

$$l(H) = -\log(|\mathbf{C_H}|) - \mathbf{X_H}^T \mathbf{C_H}^{-1} \mathbf{X_H}, \qquad (10)$$

where X_H denotes the column vector of length n of observations and C_H is a fully populated VCM, which components are given using Equation (7). $|C_H|$ is here the determinant of the matrix. By maximizing the likelihood function, one obtains an optimal choice for H: $\hat{H} = \text{argmax}(l(H))$, with $0 < H < 1$. To approximate Equation (10), matrix inversions are necessary. They can be avoided using the Whittle estimator [28], which aims to provide faster estimation with only a slight inaccuracy. In that case, the Whittle likelihood in its discretized form is given by

$$l_W(H) = -\sum_{\omega \in \Omega} \left[\log(\widetilde{f}(\omega, H)) + \frac{I(\omega)}{\widetilde{f}(\omega, H)} \right], \tag{11}$$

with Ω the set of discrete Fourier frequencies, $\widetilde{f}(\omega, H)$ the continuous-time process spectral density and $I(\omega)$ the periodogram $I(\omega) \infty \sum_{j=1}^{N} |X_{H,j} e^{-ij\omega}|^2$. The same notation as in [54] was adopted.

Whittle estimator assumes a priori that the power spectrum of the underlying process of the dataset is known. Moreover, to be applicable to fGn, the mean of the time series has to be subtracted beforehand [56]. As aforementioned, the Whittle estimator should only be used if a time-series has already been shown by other methods to be consistent with a specific process, e.g., a fGn. Thus, it is not an adequate method to detect LRD.

Wavelet Estimator (WE)

Since H describes the level of statistical self-similarity of a time series or spatial process, the exponent can be found by averaging squared values of the wavelet coefficients

$$E_j = \frac{1}{n_j} \sum_{k=1}^{n_j} |d_X(j,k)^2|, \tag{12}$$

where $d_X(j,k)$ are the detailing coefficients defined as $d_X(j,k) = \int_{-\infty}^{\infty} \psi_{j,k}(t) X_H(t) dt$, with $\psi_{j,k} = 2^{-\frac{j}{2}} \psi(2^{-j}t - k)$, ψ the mother wavelet. $X_H(t) = \sum_k \sum_{j=1}^{J} d_X(j,k) \psi_{j,k}(t) + approx$, with J, the number of decomposition level and *approx* the approximating component—not of interest for our purpose. E_j at scale j can be shown to obey the scaling law:

$$E_j \sim 2^{\alpha j}. \tag{13}$$

The Hurst exponent is obtained by fitting a line to the linear part of $\log_2(E_j)$ versus j in order to obtain the slope α. Differently to the power spectrum method, E_j contains here the information about the power carried at each time scale j. It was found to be robust even if the LRD is not equivocal [57] but performs poorly for short sample. Similarly to the power law β, α is linked to H differently for a fBm and fGn, with $H = \frac{\alpha-1}{2}, H = \frac{\alpha+1}{2}$, respectively. Wavelet based estimator are implemented in Matlab under wfbmesti. The values are based on the estimation of the Hurst exponent for a fBm and have, thus, to be applied to the cumulative sum of a fGn.

Additional Remarks

Periodicity and noise in the time series biased strongly the identification of LRD; the estimators are misleading and can detect LRD erroneously, or on the contrary find a Gaussian noise with $H = 0.5$ [20]. Frequency or wavelet-based estimators depend strongly on short-memory and necessitates strategies to alleviate these effects. The estimators have to be enabled to focus on the long-range correlation in case of additional Gaussian noise of unknown variance. One possible way to face this challenging situation will be proposed in this contribution; the filtering of the noise with a low pass Butterworth filter. Detailed simulations in Section 4 will explain the reasons of this choice.

3.3. Butterworth Filter

Butterworth filters can be designed as bandpass, lowpass, or high pass filters. They are called maximally flat filters as for a given order they have the sharpest roll-off possible without inducing peaking in the Bode plot. The Bode plot is a log–log graph where the gain in decibels is plotted against the logarithm of the angular frequency. An example is shown exemplarily in Figure 2 for different order of the Butterworth filter. We note that the Butterworth filter changes from pass band to stop-band by achieving pass band flatness. This is done at the expense of wide transition bands. This property, sometimes considered as the main disadvantage of Butterworth filter, turned out to be the main reason for using such a filter for our application. A great flexibility is given in locating the cutoff frequency, i.e., the values of the elements of the Butterworth filter are more practical and less critical than many other filter types. Interested readers should refer to [58] or [59].

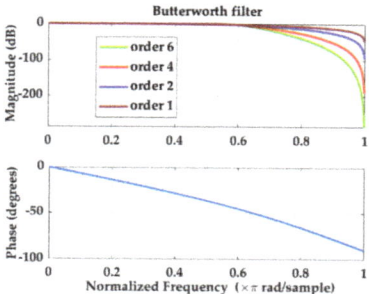

Figure 2. Bode plot for a lowpass Butterworth filter with a cutoff frequency of 300 Hz (0.6π rad/sample for data sampled at 1000 Hz). Different orders were simulated. (**bottom**): the phase response; (**top**): the magnitude.

4. Simulations and Results

In this section, we will combine all the mathematical developments presented in the previous sections: surface fitting and Hurst exponent estimation. We recall that our aim is to estimate the Hurst exponent of the range measurements from the residuals of a B-spline approximation. In order to work in a controlled framework, we use in a first step simulated TLS observations. A short real data analysis highlights the potential of the proposed methodology, which will be pursued in further dedicated contributions.

4.1. Simulation of TLS Observations

The first step towards analyzing the correlation structure of the range residuals as described in Section 2 is to simulate TLS observations. In this contribution, we choose two different surfaces with increasing complexity: a plane and a Gaussian surface.

The plane has the equation $z = -3x + 15y + 7$. The distance between the origin of the coordinates and the centre of the plane is 7 m. The coefficients of the plane were chosen without any search for optimal scanning condition in order to test our methodology in complex cases. The representation of the plane is shown in Figure 3 (left bottom). The TLS is placed at the origin of the axes, see Figure 3 (right).

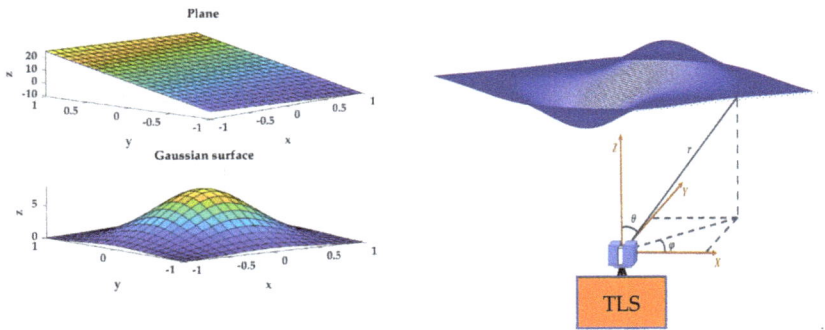

Figure 3. (**left,top**): Simulated plane. (**left,bottom**): Simulated Gaussian surface. (**right**): Origin of the laser scanner in Cartesian and polar coordinates. We call $\theta = VA, \varphi = HA$.

The Gaussian surface has the equation $z = \frac{1}{2\pi 5^2} e^{-\frac{1}{2}(\frac{x^2}{5^2} + \frac{y^2}{5^2})}$ and is shown in Figure 3 (left top).

For each surface, the PC were generated by varying $x \in [\,-1\;\;1\,], y \in [\,-1\;\;1\,]$. Two samplings were chosen: case (i) 400 observations and case (ii) 1000 observations per scanning line, resulting in PC of size 400*400 and 1000*1000, respectively. These cases are chosen to study the impact of the density of the PC on the estimation of the Hurst parameter.

4.2. Noise Simulation

The simulated Cartesian coordinates were backwards transformed into polar coordinates $[VA, HA, r]$ and noise component wise:

- to the vertical and horizontal angles is added a Gaussian noise with a standard deviation of 0.0001° generated with the Matlab function randn,
- to the range r is added a fGn noise with a standard deviation of 0.005 m. We generated noise vectors with three different Hurst exponents: 0.6 (nearly Gaussian), 0.7, and 0.9 (strong LRD).

Line Wise Noise

We did not generate one noise vector for the whole observations. Instead, we added to each scanning line an independent noise vector; see Figure 4 for an illustration of the chosen strategy. We generated as many noise vectors as scanning lines, which size depend on the chosen sampling (case (i) or (ii)).

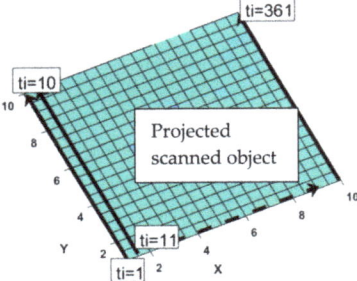

Figure 4. Explanation of the concept of line wise noise. One noise vector is added to each line for a constant x independently. In this example, from $t_i = 1$ to $t_i = 10$ is added one noise vector. A new one is added starting from $t_i = 11$. The same procedure is repeated for as many lines as the point clouds contain.

Thus, the noise is not added as a whole to the observations. We justify this line wise strategy by the fact that TLS observations are recorded in such a way that the elapsed time between the last observation of one line and the first of the following line is much longer than the time between two observations inside one line. Using this modelization, we are able to place ourselves in the context of a more general and potentially time varying correlation structure during the scanning process, answering the challenge (iii) mentioned in the introduction. This effect can be caused, e.g., by changing object properties or atmospheric conditions.

4.3. Estimation of the Hurst Exponent from the Residuals

In the following, we will compare the estimates of H with the three previously described estimators. An application to a real case scenario is presented in Section 4.5, as well as a generalization of the results in Section 4.6.

We start with the approximation of the plane (Section 4.1, Figure 3, left bottom). We approximate the PC with a cubic B-spline surface and fix the numbers of CP to estimate to 4, which is justified by the simple geometry of the simulated object [60]. Intentionally, we are not searching from an optimal functional model, which could be based on an iterative method using information criteria [37]. We take a reference value of $H_{ref} = 0.6$ for the simulated noise. This reference value is close to $H = 0.5$ (white noise) and is challenging to estimate accurately.

Interestingly, this scenario is difficult to fit with B-splines due to the unfavorable orientation of the plane in space; it leads to a so-called strong "border effect" in the B-spline surface approximation, which is not solved entirely by a higher knot multiplicity. This effect can be visually seen in the plot of the residuals by a strong increase of the variance at the beginning of each line (Figure 5 left). The correlation structure of the residuals is not dependent on the stationarity (or not) of the residual's variance. Consequently, we allowed ourselves to disregard the corresponding inaccurately approximated first epochs of each lines; exactly the same results as those presented in the next section were found. We interpret this effect as being due to the self-similarity property of the noise and, thus, we did not intend to suppress it.

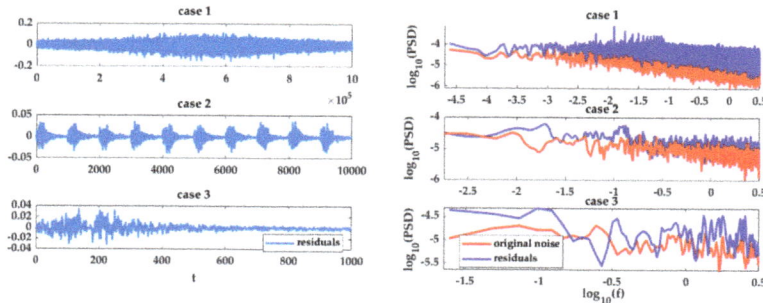

Figure 5. (left): Residuals of the plane adjustment versus time (top: case 1, the whole residuals, middle: case 2, the first 10,000 values, bottom: case 3 the first 1000 values corresponding to one scanning line). The x-axis corresponds to the time, exemplarily in (s), whereas the y-axis is the residuals, exemplarily in (m). **(right):** The corresponding PSD as a log–log plot. F is given in Hz and the PSD in dB/Hz. No additional angle noise.

4.3.1. Impact of Model Misspecification: No Noise Angle

In order to highlight the impact of both the model misspecification and the angle noise on the estimation, we firstly noised the simulated range only. On the contrary to real data analysis, the simulation framework allows for this flexibility.

The obtained range residuals of the adjustment are shown in Figure 5. They are plotted

- case 1: as a whole in Figure 5 (left top), i.e., each 1000 observations correspond to one line;
- case 2: the first 10 lines (Figure 5 left middle) and;
- case 3: only the first line in Figure 5 (left bottom).

Although the whole residuals may visually appear as white noise with a slight variance increase in the between t = 4e5 and t = 7e5 (unit of time) due to the scanning configuration (Figure 3, left top), we identify repetitive pattern for each approximated scanned line (Figure 3, left middle); they are likely to influence negatively the estimators of the Hurst parameter that are acting in the frequency domain such as the WhiE.

The PSD for the three cases (1–3) corresponding to Figure 5 (left) are plotted in Figure 5 (right). We note that it has strong similarities with the one of the simulated noise for the whole range residuals. The expected power law decay is kept nearly intact, which is beneficial to the wavelet and WhiE estimator. Additional frequencies for $-2.5 < \log_{10}(f) < 1$ are visible, which we link with the aforementioned repetitive pattern due to model misspecifications. For decreasing sample size (Figure 5, right bottom), the low frequency domain of the analyzed residuals from the first scanning line does not follow exactly the one of the original noise. It is possible to compensate for that effect using the GHE by decreasing τ_{max}; this corresponds to down-weighting the impact of the low frequencies in the estimation of H.

For the sake of convenience and without lack of generality, we will carry our explanation with the residuals of the first line.

We computed the Hurst exponent for case 1 and 3 with the chosen three methods. Case 2 is of minor interest and was only presented to show the pattern of the residuals. For case 1, the whole residuals are considered in the estimation of H, leading in a longer time series, whereas for case 3 we take the mean of the H estimated over smaller samples. For case 3, the standard deviation as well as the min/max values of H can be given (one Hurst exponent is estimated for each line). The results are presented in Table 1. We added the estimation of the Hurst exponent from the original generated noise for comparison purpose.

Table 1. Estimation of the Hurst parameters from the residuals for case 1 and case 3, for the three estimators under consideration. We give additionally the standard deviation of the estimation, when available. Case without additional angle noise, $H_{ref} = 0.6$.

	GHE	WhiE	WE
Case 1 H	0.60 (std 8×10^{-4})	0.54 (std 0.14)	0.61
Noise	0.60 (std 4×10^{-4})	0.54 (std 0.12)	0.60
Case 3 mean(H) min/max std(H)	0.61 0.51/0.72 0.03	0.7 0.5/0.9 0.14	0.60 0.35/0.72 0.07
Noise mean(H) min/max std(H)	0.59 0.55/0.65 0.02	0.7 0.5/0.9 0.14	0.60 0.37/0.70 0.06

Table 1 shows that from the three estimators, the WhiE performs worst. This holds true particularly for case 3, for which the Hurst exponent for both the simulated noise and the residuals are overestimated; this effect was expected due to the small samples under consideration (1000 observations) and is related in the literature as the main drawback of this estimator—under the assumption that the noise is fGn. For case 1 (whole residuals), the WhiE has a better performance regarding case 3 due to the frequency averaging but remains a poorer estimate compared with the values given by the GHE and the WE; both estimators provide the true H. The GHE is less affected by the sample size than the wavelet estimator, i.e., in case 3 the standard deviation of H for the WE is higher than for GHE for both the noise and the residuals.

From this simulation and without additional noise on the angles, the preference goes towards the GHE when the H exponent has to be evaluated for each line, i.e., for small samples. This is a nice result when temporal variations of the parameter are expected ([61]), since they will be detected with a higher trustworthiness than with other estimators.

Similar results are obtained for $H_{ref} = 0.7$ and $H_{ref} = 0.9$, and are not presented for the sake of shortness and readability of this article.

4.3.2. Impact of Model Misspecification and Noise Angle

In a second step, we added a noise with a standard deviation of $1\times10^{-4\circ}$ to the angle components. In order to be able to visually identify the difference between the slope of the PSD for the noise to the one of the residuals—affected by additional white noise—we consider the case $H_{ref} = 0.9$ (see Figure 6). This is a challenging exponent to estimate, since the corresponding process is close to a flicker noise. Results for other H are similar when the same methodology is applied.

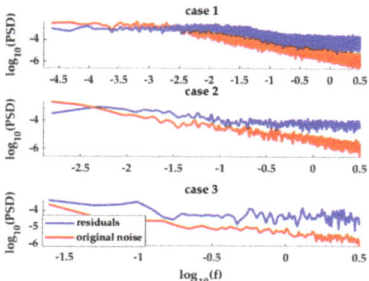

Figure 6. The PSD of the residuals for case 1 (**top**), case 2 (**middle**), and case 3 (**bottom**). The Hurst exponent for the simulated noise is $H_{ref} = 0.9$. A plane was approximated with 1000 observations per line. Log–log plot. Case with additional angle noise.

From Figure 6, the impact of the additional noise coming from the angles and propagating in the range residuals can be clearly identified in the high frequency domain, i.e., from $\log_{10}(f) > -0.6$ for case 1, and $\log_{10}(f) > -0.4$ for case 3. This corresponds to a noise at -40 dB/Hz for case 1, -45 dB/Hz for case 2, and -57 dB/Hz for case 3, approximately.

The corresponding Hurst exponents were estimated and the results are presented in Table 2. As previously, we also give the results obtained for the simulated noise.

Table 2. Estimation of the Hurst parameters from the residuals for case 1 and case 3, with the three estimators under consideration. We additionally give the standard deviation of the estimation, when available. Additional angle noise, $H_{ref} = 0.9$.

	Before Filtering			After Filtering Case 1: cutoff $\log_{10}(f_c) = -0.6$ Case 3: cutoff $\log_{10}(f_c) = -0.4$		
	GHE	WhiE	WE	GHE	WhiE	WE
Case 1 H	0.71 (std 0.01)	0.53 (std 0.3)	0.71	0.87 (std 0.01)	0.63 (std 0.3)	0.98
Noise	0.89 (std 3×10^{-4})	0.7 (std 0.2)	0.90			
Case 3 mean(H) min/max std(H)	0.71 0.56/0.86 0.05	0.67 0.50/0.96 0.15	0.71 0.51/0.92 0.08	0.89 0.76/0.96 0.03	3 1.52/4.58 1.51	0.96
Noise mean(H) min/max std(H)	0.86 0.75/0.95 0.03	0.64 0.51/0.75 0.20	0.83 0.74/0.98 0.08			

The first remark to draw from Table 2 is the stronger difficulty to estimate the Hurst parameter from the true fGn for small samples (case 3) than for longer sample (case 1). The mean values are slightly below the true one of $H_{ref} = 0.9$ for the GHE and WE estimators, with a higher standard deviation than in the previous case with $H_{ref} = 0.6$. Clearly, the WhiE performs poorly and systematically underestimates the true parameter.

The second remark is the impossibility to extract the correct, or a value close to the correct Hurst exponent, independently of the case under consideration. The noise of the angles, as well as the noise induced by the fitting, leads to a strong underestimation of H close to 0.7. The decrease towards $H = \frac{1}{2}$ (a white noise) is due to the increase of white noise in the signal. As previously, the WhiE estimates poorly H (0.53 for case 1). It is shown to be thus strongly affected by additional white noise on the residuals.

The analysis of the PSD (Figure 7) for case 1 and 3 highlights the impact of additional white noise. From Equation (9) and Equation (13), we notice that the GHE and WE need both low and high frequencies to perform the approximation of the Hurst exponent with trustworthiness; it seems advantageous to filter the high frequency noise of the residuals. In this contribution, we propose to apply a lowpass Butterworth filter of first order on the residuals from the cutoff frequency at which the PSD kicks towards white noise. This choice is not justified by empirical findings and we propose in the following to detail the reasons why we opted for the Butterworth filter.

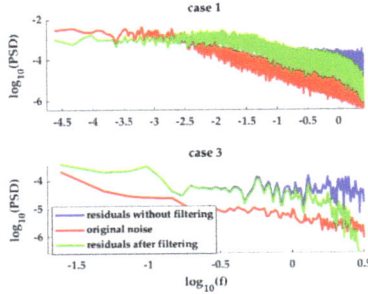

Figure 7. The PSD of the residuals for case 1 (**top**) and case 3 (**bottom**). The Hurst exponent for the simulated noise is $H_{ref} = 0.9$. A plane is approximated with 1000 observations per line. Case with additional angle noise. The red curve corresponds to the PSD of the simulated noise, the blue curve to the PSD of the residuals and the green to the filtered residuals.

Why a Butterworth Filter

A sharp low pass filtering would lead to an abrupt decrease of the PSD from the cutoff frequency of the filter; this effect is here unwanted as the estimation of H necessitates the whole range of frequencies, which would not be given any more. We prefer, thus, the "smooth and gentle" Butterworth filter of first order; it allows a continuous decrease of the high frequencies from the cutoff frequency. This leads to a filtering shown Figure 7 (green line), where the PSD of the filtered signal has the same decrease as the reference noise from $\log_{10}(f) = -2$ for case 1 and $\log_{10}(f) = 0$ for case 3. The filtering leads to an estimate of the Hurst exponent of 0.87 (std 0.01) and 0.89 (std 0.03) with the GHE for case 1 and 3, respectively (see Table 2). τ_{max} was fixed to 20, as proposed in the literature [51]. Increasing τ_{max} leads to a slight decrease of H of 0.2 for $\tau_{max} = 40$ with an increase of the std to 0.03, whereas $\tau_{max} = 5$ is linked with an increase of H of 0.3 by a decrease of the std to 0.008. Thus, a balance has to be found to fix τ_{max} optimally. A deep analysis of the PSD, i.e., the amount of power at low frequencies is necessary; whereas τ_{max} can be used to filter unwanted low frequencies due to model misspecification, the cutoff frequency acts on the high frequency domain of the PSD.

With the chosen cutoff frequency, the WE overestimates H. Using a cutoff frequency of $\log_{10}(f_c) = -0.35$, instead of $\log_{10}(f_c) = 0.4$, for case 3 yields $H = 0.88$. Similar results are obtained for case 1 with the WE by increasing $\log_{10}(f_c)$ to -0.55. Unfortunately, the GHE decreases to 0.83 in both cases. However, considering that (1) no prior knowledge of the Hurst exponent was available, (2) additional white noise affects the residuals, and (3) model misspecification are present, this remains a good approximation of the true H of 0.9. Indeed, this value of the Hurst parameter is known to be challenging to estimate since it is close to the limit between fGn and fBm.

Why First Order?

The answer is strongly related to the previous one: as shown in Figure 2 from the Bode plot, the flatness of the filter is of main importance to ensure smooth transition in the PSD of the filtered signal.

4.3.3. Sensitivity Analysis: Impact of the Cutoff Frequency

In this section, we propose to analyze the sensitivity of the estimated H exponent regarding the chosen cutoff frequency. Figure 8 summarizes the results for case 1 (top) and 3 (bottom) by varying $\log_{10}(f)$ from to -0.7 to -0.25.

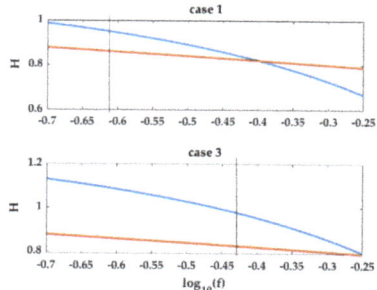

Figure 8. Sensitivity analysis of the estimated H from the residuals of the B-spline surface fitting by varying the cutoff frequency. A plane is estimated with 1000 observations per line and the simulated noise is fGn with $H_{ref} = 0.9$. Case 1 corresponds to the whole residuals, case 3 to the first line. Case with additional angle noise. The red curve corresponds to the GHE, the blue curve to the WE.

With great evidence, the GHE is much less sensitive to the cutoff frequency than the WE. A linear dependency can be found, with a variation of H from 0.88 to 0.78 for the chosen range of cutoff frequencies and from 0.87 to 0.80 for case 1 and 3, respectively. H, estimated with WE, has a much higher range of values—from 1 to 0.68 and to 1.15 to 0.8 for case 1 and 3, respectively. From these results, and considering that we placed ourselves intentionally in a challenging estimating scenario with a strong Hurst exponent, we recommend using the GHE instead of the WE when the residuals are filtered and small samples are considered.

4.3.4. Small Samples

In this section, we place ourselves in case (i) as described in Section 4.1. and generate smaller samples of 400 observations per line. We chose three values for H: 0.6, 0.7, and 0.9, and apply our methodology to filter the residuals from additional white noise and/or results from model misspecifications. We identify the cutoff frequency f_c visually by plotting (1) the PSD of the whole residuals for case 1 and (2) the PSD of 10 randomly chosen lines for case 3 and averaging the identified cutoff frequencies. The Hurst exponent is estimated with the GHE; due to their asymptotic properties, the WE and the WhiE are known to perform poorly for small samples [62].

For $H_{ref} = 0.6$, the PSD is nearly similar to a white noise (see Figure 1). This leads to a stronger difficulty to identify the PSD kick. Nevertheless, we were able to identify with a high confidence the correct cutoff frequency and a value of $H = 0.60$ could be estimated for case 1 for $\log_{10}(f_c) = -0.5$. We link this result with the use of a Butterworth filter of the first order and the low sensitivity of the GHE to a misspecification of the cutoff frequency. The same cutoff was used for the two other simulated H_{ref}. This result strongly confirms the feasibility of extracting the Hurst exponent from residuals of regression B-splines, in the presence of both functional misspecifications and additional unknown noise. The cutoff frequency depends on the noise angle variance, as illustrated in Table 3. Increasing the $\sigma_{HA} = \sigma_{VA}$ to $0.001°$ instead of $0.0001°$ yields a different cutoff frequency. We intentionally do not present this result in order not to overload the readers with simulation results that lead to similar conclusions.

Table 3. Estimation of the Hurst parameters from the residuals for case 1 and case 3, with the GHE and with additional angle noise for two standard deviations ($1 \times 10^{-4°}$ and $1 \times 10^{-3°}$). $H_{ref} = 0.6, 0.7,$ and 0.9. The cutoff frequencies (f_c) are visually determined.

	Std Noise Angle $1 \times 10^{-4°}$		Std Noise Angle $1 \times 10^{-3°}$	
	Case 1 Cutoff $\log_{10}(f) = -0.5$	Case 3 Cutoff $\log_{10}(f) = -0.25$	Case 1	Case 3
$H_{ref} = 0.6$	0.60 (std 0.05)	0.63 (std 0.03)		
$H_{ref} = 0.7$	0.68 (std 0.02)	0.70 (std 0.03)	$\log_{10}(f_c) = -0.5$ 0.66 (std 0.08) $\log_{10}(f_c) = -0.4$ 0.71 (std 0.07)	$\log_{10}(f_c) = -0.25$ 0.66 (std 0.03) $\log_{10}(f_c) = -0.55$ 0.72 (std 0.07)
$H_{ref} = 0.9$	0.88 (std 0.02)	0.87 (std 0.04)		

4.4. Result for Gaussian Surface

The second example corresponds to a simulated Gaussian surface (case (ii), Section 4.1.). Ten CP in the two directions were estimated with B-splines of order three. The stationary reference noise was simulated with $H_{ref} = 0.7$ and 1000 observations per line. Similarly to the previous simulations, we do not aim to optimally fit the surfaces so that the impact of potential misspecification can be considered. In Figure 9 (left), the PSD of the residuals together with the PSD of the simulated noise are shown; Figure 9 (right) represents the residuals for case 1 and 3 respectively, following the previous section. This latter figure highlights the lack of repetitive patterns in the residuals plotted per line (case 3). Only a steady increase of the variance towards the middle of the surface can be seen, which is coherent with the Gaussian form of the surface (Figure 2 left bottom). This behavior does not affect the estimation of the Hurst parameter, which was 0.72 (std 1×10^{-3}) for case 1 and 0.71 (std 0.01) for case 3 with or without filtering. From the PSD, a low additional white noise from $\log_{10}(f_c) = 0$ could be identified, which did not affect the determination of H. We interpret this lack of additional white noise in the residuals as coming from the goodness of the surface approximation, i.e., the B-splines themselves are acting as a low pass filter so that no additional noise coming from the angles could drift into the residuals in that case. However, we were able to decrease the estimated Hurst parameter for case 1 to 0.69 (std 5×10^{-4}) by decreasing τ_{max} to 10, i.e., decreasing the impact of the low frequencies. In this case, the B-spline LS system filters the low frequencies domain strongly, which could be accounted for by acting on τ_{max}.

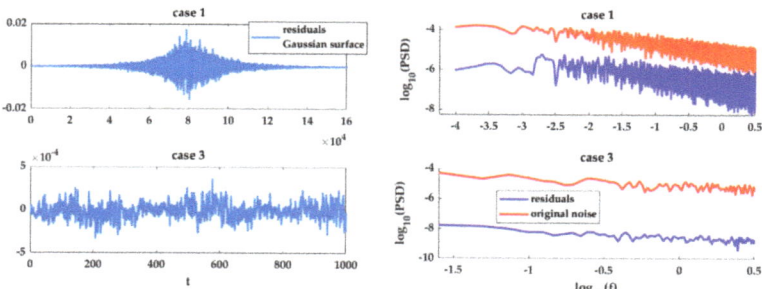

Figure 9. (**left**): Residuals of the B-spline approximation for a Gaussian surface. Case 1 (**top**); the whole residuals and case 3: the 1000 first observations corresponding to one scanning line. (**right**): The corresponding PSD (red the original noise, blue the residuals).

This correction highlights the potential of our methodology to identify and filter model misspecification from the LS residuals. It is and remains based on a visual analysis of the PSD and an understanding of the residuals as prior to the estimation of the Hurst parameter. It is not recommended to use a bandpass Butterworth (or any other filter such as a notch filter) to filter specific frequencies. This was shown to strongly affect the determination of the Hurst exponent by creating an artificial decrease of frequencies amplitude in the middle of the frequency range, where a regular decrease is of main importance for the determination of H.

4.5. Application to Real Data

We propose to apply the proposed methodology to a real case scenario. Unfortunately, the true correlation structure is unknown; the development of a model based on a physical explanation of the TLS correlation is beyond the scope of this paper and led to further studies.

A white plane of size 1 m*1 m was scanned at a distance of 10 m with a Z+F 2016F using the scanning modus "extremely high", with which 1 Mio. per s can be recorded. The scanning configuration is presented in Figure 10 (left); it is optimal and corresponds to the simulated data with no tilt and the TLS pointing in the direction of the z-axis. The obtained point cloud was pre-processed to avoid edge effects and outliers, and cut using a free software. We finally approximated the data with a cubic B-spline, following the methodology presented in Section 2. The residuals for one scanning line are plotted in Figure 10 (right, top), together with the corresponding PSD (Figure 10, right, bottom, blue line). We visually identified a cutoff frequency of $\log_{10}(f_c) = 0$, which we used to filter the residuals with a Butterworth filter (Figure 10, bottom, yellow line). As for the simulations, the results obtained with the three Hurst estimators proposed in this contribution differ. Without filtering, we found values of 0.85 for the GHE with $\tau_{max} = 20$ (which was chosen due to the lack of additional low frequencies from inaccurate functional model), 1.01 (i.e., flicker noise) for the WhiE, and 0.61 for the WE. This last result highlights that the WE is affected by white noise—the value found was close to 0.5—and by the small number of observations used (900 per line). The tendency for the WhiE to overestimate the Hurst parameter with respect to the GHE (Table 1) is additionally shown. Using the visually identified cutoff frequency, the Hurst estimator was increased to 0.88 for the GHE, but stayed constant for the WE; this estimator is definitively not a relevant choice for the case study under consideration. A high value of 1.61 was found for the WhiE, which seems not usable with the filtered residuals, i.e., the WhiE being a spectral estimator is affected by the strong decrease of the PSD at high frequencies.

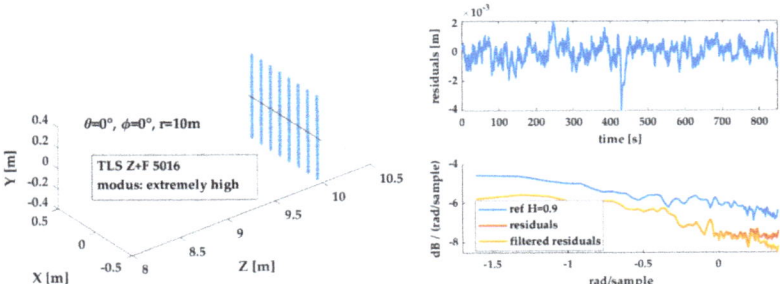

Figure 10. (**left**): The schematic scanning configuration: no tilt and a distance of 10 m to the center of the coordinate system. Right, top: residuals of the B-spline approximation for the plane under consideration. (**right,bottom**): The corresponding PSD. The blue line is a reference fGn with $H_{ref} = 0.9$, the red line corresponds to the residuals and the yellow one to the filtered residuals with $\log_{10}(f_c) = 0$.

From Figure 10 (right bottom) the plausibility of considering the noise of the TLS range as being fGn is confirmed; the blue line corresponds to a reference fGn of 0.9 and is nearly parallel to the yellow one from the filtered residuals.

This short case study validates the proposed methodology for a real case scenario: it is feasible to estimate the Hurst parameter from the range residuals of a plane scanned with a TLS. Further studies will be carried out in a next future to investigate more deeply the correlation structure. This latter is expected to depend on, e.g., the scanning rate, distance from the plane to the TLS, or atmospheric conditions.

4.6. Summary: A Methodology to Extract the Hurst Parameter from the B-Spline Residuals

In this section, we summarize our methodology to extract the Hurst parameter of the underlying noise from TLS range measurements from LS residuals (Figure 11). We recall that the noise is simulated line wise as a fGn with a given Hurst parameter varying from 0.5–1 (persistent correlations). Working with real observations this assumption has to be tested by analyzing the stationarity of the time series as well as the power law of its PSD.

We start with the raw polar observations, which are to be transformed into Cartesian coordinates. After having parametrized the point cloud, a B-spline surface approximation is performed. The choice of the order of the B-splines is left to the user (e.g., cubic B-splines), as well as the method to fix the knot vector optimally or/and the number of CP to estimate. The residuals of the approximation are transformed backwards into polar coordinates; only the range residuals are further analyzed. They are plotted as a whole and line wise to visually identify the potential impact of model misspecification (low frequencies, repetitive pattern). These patterns could act—as a snow ball effect—on the determination of the Hurst parameter. As an important tool to understand the structure of the residuals, the PSD is plotted against the frequencies (a log–log plot should be used for a better visualization). Additional white noise or model misspecification can slide into the frequency domain; they are identified and filtered with a low pass Butterworth filter of first order. We recommend the use of the generalized Hurst estimator. This latter was shown to be robust to slight uncertainties in the determination of the cutoff frequency, as well as less sensitive to small samples effect, compared with the wavelet estimator. Thus, temporal variations of the Hurst exponent can be analyzed by making a line wise analysis of the Hurst parameters.

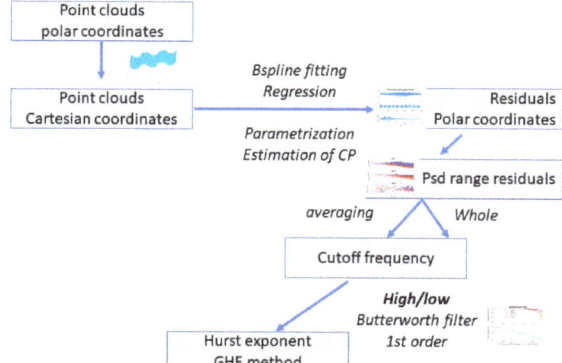

Figure 11. Summary of the methodology to extract the Hurst parameter from the range residuals of a B-spline approximation from a TLS point cloud.

5. Conclusions

In this contribution, we have developed and validated an innovative yet simple strategy to extract the correlation structure of the underlying observation noise from the residuals of a B-spline surface approximation. This determination is neither based on least-squares estimation or collocation, nor parametric, and has the main advantages of being easy to use and computationally efficient.

Our case study dealt with TLS raw observations, having in mind to analyze the correlations of the range observations to perform more rigorous and trustworthy statistical tests for deformation of scanned objects. This is a highly relevant application for avoiding and/or quantifying the potential risk related to the deformation of structures such as dams or bridges. Moreover, knowledge of the correlation structure could serve to predict future deformations.

The range measurements of TLS observations are known to be temporally correlated. We guess from physical consideration that the power spectral density of the noise could be represented by a power law. The framework of LRD allows description of such kinds of noise accurately. In this study, we chose to model the correlation structure of the TLS range measurements by a stationary persistent fGn. The fGn is widely used to describe all kinds of noise in various domains and can be fully described by means of its Hurst exponent (related to the fractal dimension) and its variance. There exist various estimators for the Hurst exponent. In this contribution, we compared the performance of one of each family: the generalized Hurst estimator, the Whittle likelihood estimator and the wavelet estimator. We simulated small and longer samples, as well as observations noise with different Hurst exponents. Our goal was to determine as accurately as possible, from the B-splines range residuals, the reference parameter. Regression B-spline surface fitting can be applied to nearly every noisy and scattered point cloud, without limitation to specific surfaces such as circle or plane. Even if they are structurally correlated, the residuals of the approximation still contain information about the correlation and noise structure of the raw observations.

Unfortunately, as in every approximation model, misspecifications are likely to arise. They introduce additional frequencies in the residuals, which affect the determination of the Hurst parameter. Simulating a plane, we identified unwanted white noise as strongly affecting the estimation. A low pass Butterworth filter of the first order applied to the residuals was able to correct the bias induced by an unwanted additional white noise. The generalized Hurst estimator was shown to be robust against slight over or underestimation of the cutoff frequency of the filter. The Whittle likelihood performs badly in estimating H, which was linked to the potential non-stationarity of the residuals, i.e., the assumption that the residuals should be fGn is mandatory for this estimator. The wavelet estimator performs ideally in absence of white noise and could be shown to be sensitive to the choice of the cutoff frequency. We interpreted this behavior as being linked with non-averaging, compared with the GHE.

Simulating a Gaussian surface, the impact of model misspecification in the low frequency domain was highlighted and filtered adequately with a high pass Butterworth filter to improve the determination of the Hurst exponent.

For both simulated cases, similar conclusions were drawn; the Hurst exponent can be well determined with the GHE, provided that a prefiltering of the residuals with the smooth Butterworth filter of first order is performed. The cutoff frequency could be visually identified from the PSD of the residuals (line wise or as a whole). The feasibility of the proposed methodology was confirmed using real data from a plane scanned with a TLS with the "extremely high" resolution.

This powerful way to identify the noise structure from the residuals paves the way for a deeper study of the correlation dependency of TLS range measurements, independent of specific calibration procedures. Due to the high accuracy and precision of the determination of the fractal parameter, potential atmospheric parameters could be deduced from the B-spline residuals, as well as sensor characteristics. This analysis will be the topic of a later study with real data. The estimation of the range variance remains to be solved. A proposal could be based on the calibration of the LS system with white noise.

Funding: The publication of this article was funded by the Open Access fund of Leibniz Universität Hannover. The author gratefully acknowledge the funding by the Deutsche Forschungsgemeinschaft under the label KE 2453/2-1.

Conflicts of Interest: The author declares no conflict of interest.

References

1. Vosselman, G.; Maas, H.G. *Airborne and Terrestrial Laser Scanning*; CRC Press: Boca Raton, FL, USA, 2010.
2. Pelzer, H. *Zur Analyse Geodätischer Deformationsmessungen*; Verlag der Bayer. Akad. d. Wiss: Munchen, Germany, 1971.
3. Lee, S.Y.; Wolberg, G.; Shin, S.Y. Scattered data interpolation with multilevel B-splines. *IEEE Trans. Vis. Comput. Graph.* **1997**, *3*, 228–244. [CrossRef]
4. Aigner, M.; Jüttler, B. Distance regression by Gauss–Newton-type methods and iteratively re-weighted least-squares. *Computing* **2009**, *86*, 73–87. [CrossRef]
5. Koch, K.R. Fitting free-form surfaces to laserscan data by NURBS. *AVN Allg. Vermess.-Nachr.* **2009**, *116*, 134–140.
6. Kermarrec, G.; Neumann, I.; Alkhatib, H.; Schön, S. The stochastic model for Global Navigation Satellite Systems and terrestrial laser scanning observations: A proposal to account for correlations in least squares adjustment. *J. Appl. Geod.* **2019**, *13*, 93–104. [CrossRef]
7. Wheelon, A.D. *Electromagnetic Scintillation: Part I Geometrical Optics*; Cambridge University Press: Cambridge, UK, 2001.
8. Ishimaru, A. *Wave Propagation and Scattering in Random Media*; IEEE Press and Oxford University Press: New York, NY, USA, 1997.
9. Kauker, S.; Holst, C.; Schwieger, V.; Kuhlmann, H.; Schön, S. Spatio-temporal correlations of terrestrial laser scanning. *AVN Allg. Vermess.-Nachr.* **2016**, *6*, 170–182.
10. Kauker, S.; Schwieger, V. A synthetic covariance matrix for monitoring by terrestrial laser scanning. *J. Appl. Geod.* **2017**, *11*, 77–87. [CrossRef]
11. Koch, K.-R. *Parameter Estimation and Hypothesis Testing in Linear Models*; Springer: Berlin, Germany, 1999.
12. Stein, M.L. *Interpolation of Spatial Data: Some Theory for Kriging*; Springer: New York, NY, USA, 1999.
13. Montillet, J.-P.; Bos, M.S. *Geodetic Time Series Analysis in Earth Sciences*; Springer: Cham, Switzerland, 2020.
14. Mandelbrot, B.B.; Van Ness, J.W. Fractional Brownian motion, fractional noises and applications. *SIAM Rev.* **1968**, *10*, 422–437. [CrossRef]
15. Sims, D.W.; Southall, E.J.; Humphries, N.; Hays, G.C.; Bradshaw, C.J.A.; Pitchford, J.W.; James, A.; Ahmed, M.Z.; Brierley, A.; Hindell, M.A.; et al. Scaling laws of marine predator search behaviour. *Nature* **2008**, *451*, 1098–1102. [CrossRef] [PubMed]
16. Keshner, M.S. 1/f noise. *Proc. IEEE* **1982**, *70*, 212–218. [CrossRef]
17. Accardo, A.; Affinito, M.; Carrozzi, M.; Bouquet, F. Use of the fractal dimension for the analysis of electroencephalographic time series. *Biol. Cybern.* **1997**, *77*, 339–350.

18. Vandewalle, N.; Ausloos, M. Coherent and random sequences in financial fluctuations. *Phys. A: Stat. Mech. Appl.* **1997**, *246*, 254–459. [CrossRef]
19. Abry, P.; Sellan, F. The wavelet-based synthesis for fractional Brownian motion proposed by F. Sellan and Y. Meyer: remarks and fast implementation. *Appl. Comput. Harmon. Anal.* **1996**, *3*, 377–383.
20. Bardet, J.-M.; Lang, G.; Oppenheim, G.; Philippe, A.; Stoev, S.; Taqqu, M.S. Semi-parametric estimation of the long-range dependence parameter: A survey. In *Theory and Applications of Long-Range Dependence*; Doukhan, P., Oppenheim, G., Taqqu, M., Eds.; Springer: Berlin/Heidelberg, Germany, 2003; pp. 557–577.
21. Taqqu, M.S. Fractional Brownian motion and long-range dependence. In *Theory and Applications of Long-Range Dependence*; Doukhan, P., Oppenheim, G., Taqqu, M., Eds.; Birkhäuser: Boston, MA, USA, 2001.
22. Stolojescu, C.; Isar, A. A comparison of some Hurst parameter estimators. In Proceedings of the 2012 13th International Conference on Optimization of Electrical and Electronic Equipment (OPTIM), Brasov, Romania, 24–26 May 2012; pp. 1152–1157.
23. Krakovská, H.; Krakovská, A. Fractal Dimension of Self-Affine Signals: Four Methods of Estimation. *arXiv* **2016**, arXiv:1611.06190.
24. Cannon, M.J.; Percival, D.B.; Caccia, D.C.; Raymond, G.M.; Bassingthwaighte, J.B. Evaluating scaled windowed variance methods for estimating the Hurst coefficient of time series. *Physica A* **1997**, *241*, 606–626. [CrossRef]
25. Hurst, H.E. Long-term storage capacity of reservoirs. *Trans. Am. Soc. Civ. Eng.* **1951**, *116*, 770.
26. Peng, C.K.; Buldyrev, S.V.; Havlin, S.; Simons, M.; Stanley, H.E.; Goldberger, A.L. Mosaic organization of DNA nucleotides. *Phys. Rev. E* **1994**, *49*, 1685–1689. [CrossRef]
27. Geweke, J.; Porter-Hudak, S. The estimation and application of long memory time series models. *J. Time Ser. Anal.* **1983**, *4*, 221–238. [CrossRef]
28. Whittle, P. Estimation and information in stationary time series. *Ark. Mat.* **1953**, *2*, 423–434. [CrossRef]
29. Abry, P.; Flandrin, P.; Taqqu, M.S.; Veitch, D. Self-similarity and long-range dependence through the wavelet lens. In *Theory and Applications of Long-Range Dependence*; Doukhan, P., Oppenheim, G., Taqqu, M., Eds.; Birkhäuser: Boston, MA, USA, 2003; pp. 527–556.
30. Kermarrec, G.; Alkhatib, H.; Bureick, J.; Kargoll, B. Impact of mathematical correlations on the statistic of the congruency test case study: B-splines surface approximation from bridge observations. In Proceedings of the 4th Joint International Symposium on Deformation Monitoring (JISDM), Athens, Greece, 15–17 May 2019.
31. Bureick, J.; Alkhatib, H.; Neumann, I. Robust spatial approximation of laser scanner points clouds by means of free-form curve approaches in deformation analysis. *J. Appl. Geod.* **2016**, *10*, 27–35. [CrossRef]
32. De Boor, C.A. *Practical Guide to Splines*; Revised ed.; Springer: New York, NY, USA, 2001.
33. Piegl, L.; Tiller, W. *The NURBS Book*; Springer Science & Business Media: Berlin, Germany, 1997.
34. Bracco, C.; Giannelli, C.; Sestini, A. Adaptive scattered data fitting by extension of local approximations to hierarchical splines. *Comput. Aided Geom Des* **2017**, *52–53*, 90–105. [CrossRef]
35. Zhao, X.; Kermarrec, G.; Kargoll, B.; Alkhatib, H.; Neumann, I. Influence of the simplified stochastic model of TLS measurements on geometry-based deformation analysis. *J. Appl. Geod.* **2019**, *13*, 199–214. [CrossRef]
36. Teunissen, P.J.G. *Testing Theory: An Introduction*; VSSD Publishing: Delft, The Netherlands, 2000.
37. Alkhatib, H.; Kargoll, B.; Bureick, J.; Paffenholz, J.A. Statistical evaluation of the B-Splines approximation of 3D point clouds. In Proceedings of the FIG-Kongresses, Istanbul, Türkey, 6–11 May 2018.
38. Soudarissanane, S.; Lindenbergh, R.; Menenti, M.; Teunissen, P. Scanning geometry: Influencing factor on the quality of terrestrial laser scanning points. *ISPRS* **2011**, *66*, 389–399. [CrossRef]
39. Boehler, W.; Marbs, A. 3D Scanning instruments. In Proceedings of the CIPA WG6 International Workshop on Scanning for Cultural Heritage Recording, Corfu, Greece, 1–2 September 2002.
40. Zhao, X.; Alkhatib, H.; Kargoll, B.; Neumann, I. Statistical evaluation of the influence of the uncertainty budget on B-spline curve approximation. *J. Appl. Geod.* **2017**, *11*, 215–230. [CrossRef]
41. Wujanz, D.; Burger, M.; Mettenleiter, M.; Neitzel, F. An intensity-based stochastic model for terrestrial laser scanners. *ISPRS J. Photogramm. Remote Sens.* **2017**, *125*, 146–155. [CrossRef]
42. Wujanz, D.; Burger, M.; Tschirschwitz, F.; Nietzschmann, T.; Neitzel, F.; Kersten, T.P. Determination of intensity-based stochastic models for terrestrial laser scanners utilising 3D-point clouds. *Sensors* **2018**, *18*, 2187. [CrossRef] [PubMed]
43. Beran, J. *Statistics for Long Memory Processes*; Chapman and Hall: New York, NY, USA, 1994.
44. Mandelbrot, B.B. *The Fractional Geometry of Nature*; W.H. Freeman: New York, NY, USA, 1983.

45. Eke, A.; Hermán, P.; Bassingthwaighte, J.; Raymond, G.; Percival, D.B.; Cannon, M.J.; Ikrényi, C.; Balla, I. Physiological time series: distinguishing fractal noises from motions. *Pflügers Arch—Eur. J. Physiol.* **2000**, *439*, 403–415. [CrossRef] [PubMed]
46. Abry, P.; Veitch, D. Wavelet analysis of long-range dependent traffic. *IEEE Trans. Inf. Theory* **1998**, *44*, 2–15. [CrossRef]
47. Perrin, E.; Harba, R.; Jennane, R.; Iribarren, I. Fast and exact synthesis for 1-D fractional Brownian motion and fractional Gaussian noises. *IEEE Signal Process. Lett.* **2002**, *9*, 382–384. [CrossRef]
48. Liu, Y.; Liu, Y.; Wang, K.; Jiang, T.; Yang, L. Modified periodogram method for estimating the Hurst exponent of fractional Gaussian noise. *Phys. Rev. E* **2009**, *80*, 066207. [CrossRef]
49. Jeong, H.D.J.; Lee, J.S.R.; McNickle, D.; Pawlikowski, P. Distributed Steady-State Simulation of Telecommunication Networks with Self-Similar Teletraffic. *Simul. Model. Pract. Theory* **2005**, *13*, 233–256. [CrossRef]
50. Barabasi, A.L. Vicsek Multifractality of self-affine fractals. *Phys. Rev. A* **1991**, *44*, 2730–2733. [CrossRef] [PubMed]
51. Sensoy, A. Generalized Hurst Exponent approach to efficiency in MENA markets. *Phys. Rev A* **2013**, *392*, 5019–5026. [CrossRef]
52. Taqq, M.S.; Teverovsky, V. On estimating the intensity of long-range dependence in finite and infinite variance time series. In *A Practical Guide to Heavy Tails*; Feldman, R.E., Afler, R.J., Taqqu, M.S., Eds.; Birkhäuser: Boston, MA, USA, 1998; pp. 177–217.
53. Di Matteo, T.; Aste, T.; Dacorogna, M.M. Long-term memories of developed and emerging markets: Using the scaling analysis to characterize their stage of development. *J. Bank. Finance* **2005**, *29*, 827–851. [CrossRef]
54. Sykulski, A.M.; Olhede, S.C.; Guillaumin, A.P.; Lilly, J.M.; Early, J.J. The debiased Whittle likelihood. *Biometrika* **2019**, *106*, 251–266. [CrossRef]
55. Brockwell, P.J.; Davis, R.A. *Time Series: Theory and Methods*; Springer: New York, NY, USA, 1991.
56. Chang, Y.C. Efficiently Implementing the Maximum Likelihood Estimator for Hurst Exponent. *Math. Probl. Eng.* **2014**, *2014*, 490568. [CrossRef]
57. Tarnopolski, M. On the relationship between the Hurst exponent, the ratio of the mean square successive difference to the variance, and the number of turning points. *Phys. A: Stat. Mech. Appl.* **2016**, *461*, 662–673. [CrossRef]
58. Butterworth, S. On the Theory of Filter Amplifiers. *Wirel. Eng.* **1930**, *7*, 536–541.
59. Zumbahlen, H. CHAPTER 8—Analog Filters. In *Linear Circuit Design Handbook*; Zumbahlen, H., Ed.; Newnes: New South Wales, Australia, 2008; pp. 581–679.
60. Kermarrec, G.; Alkhatib, H.; Neumann, I. On the Sensitivity of the Parameters of the Intensity-Based Stochastic Model for Terrestrial Laser Scanner. Case Study: B-Spline Approximation. *Sensors* **2018**, *18*, 2964. [CrossRef]
61. Garcin, M. Estimation of time-dependent Hurst exponents with variational smoothing and application to forecasting foreign exchange rates. *Phys. A: Stat. Mech. Appl.* **2017**, *483*, 462–479. [CrossRef]
62. Kirichenko, L.; Radivilova, T.; Deineko, Z. Comparative analysis for estimating of the Hurst exponent for stationary and nonstationary time series. *Inf. Technol. Knowl.* **2011**, *5*, 371–388.

© 2020 by the author. Licensee MDPI, Basel, Switzerland. This article is an open access article distributed under the terms and conditions of the Creative Commons Attribution (CC BY) license (http://creativecommons.org/licenses/by/4.0/).

MDPI
St. Alban-Anlage 66
4052 Basel
Switzerland
Tel. +41 61 683 77 34
Fax +41 61 302 89 18
www.mdpi.com

Mathematics Editorial Office
E-mail: mathematics@mdpi.com
www.mdpi.com/journal/mathematics

www.ingramcontent.com/pod-product-compliance
Lightning Source LLC
LaVergne TN
LVHW070724100526
838202LV00013B/1162